SPEAK to the EARTH
and
IT WILL TEACH YOU

The Life and Times of
EARL DOUGLASS
1862-1931

G.E. DOUGLASS

Copyright © 2009 G.E. Douglass
All rights reserved.

ISBN: 1-4392-4437-5
ISBN-13: 9781439244371

Visit www.booksurge.com to order additional copies.

CONTENTS

Introduction vii
Acknowledgements ix
Preface by G. E. Douglass xiii

List of chapters
1. EARL DOUGLASS: A Heritage of Tenacity and Love of Freedom 1
2. A COUNTRY SCHOOL TEACHER: Facing His Dilemmas 7
3. EARLY SCIENTIFIC INTEREST: A Love Affair 17
4. HOMETOWN TEACHER: Humiliation 23
5. A STEP WESTWARD: The Prairies of South Dakota 27
6. A START IN COLLEGE: The University of South Dakota 37
7. PEDDLING A BOOK: A Sore Trial 41
8. BROOKING: South Dakota Agricultural College 49
9. BOTANIST: Mexico and St. Louis 55
10. COLLEGE UPRISING: Expelled 61
11. AMES, IOWA: A Degree at Last 73
12. WESTWARD BOUND: Montana Fossil Beds 79
13. AN IDEAL COUNTRY SCHOOL: Fossils at Last 85
14. A GROWING ENTHUSIASM: Attacking a Fossil Skull 95
15. THE RUBY VALLEY: New Scenery 103
16. EXPEDITION OF 1897: Yellowstone Park 111
17. EXPEDITION OF 1898: The Bitterroot Valley 125
18. A MASTERS DEGREE: University of Montana, Missoula 135
19. EXPEDITION OF 1900: A Newspaper Helps 143
20. PRINCETON UNIVERSITY: A Storehouse of Scientific Knowledge 157
21. PRINCETON EXPEDITION 1901: A Remarkable Discovery 165

22. CONTRIBUTIONS TO SCIENCE: A Position with Carnegie 175
23. EXPEDITION OF 1903: A Troubled Love Affair 183
24. PITTSBURGH, PENNSYLVANIA: Preparing a Home............ 197
25. CARNEGIE MUSEUM: Founder's Day 1907.................... 207
PHOTOGRAPH SECTION **220**
26. THE WEST BECKONS AGAIN: The Land of the Telmathere.... 253
27. EXPEDITION OF 1909: Carnegie Wanted Dinosaurs.......... 261
28. A DISCOVERY: A Vision, a Mirage? 277
29. DIGGING OUT DINOSAURS: A Cold Winter 299
30. A DREAM, COMES TRUE: An Outstanding Discovery 317
31. CAPTURED BY THE WILDERNESS: Dreams of a Ranch........ 339
32. MY BOYHOOD DAYS: Early Memories at Dinosaur........... 353
33. OUR HOME IN THE WILDERNESS: Developing a Ranch 361
34. FINANCIAL DISASTER: A Lost Dream........................369
35. A NATIONAL MONUMENT: Dreams of a Natural Museum 375
36. A LAND OF UNDEVELOPED RESOURCES: There Must be Oil .387
37. PHILOSOPHY 399

TABLE OF PHOTOGRAPHS

Coverpage – Earl Douglass hosting visitors at the discovery site of the Carnegie Quarry (aka Dinosaur National Monument) August 1909 220

Frontispiece – Map of Carnegie Quarry area, circa 1908 - 1920. Landowners mentioned in this book 221

PHOTOGRAPHS

1   The Douglass family about 1890: Abigail, Nettie, Ida, Fernando and Earl 222
2   Portrait of Earl Douglass in St. Louis – 1900 223
3   Three generations of Douglass males: Fernando, Gawin, and Earl 223

4	Brookings Institute – "Truth or Bust" group	224
5	Earl reading under the Lone Tree by the Green River	224
6	Gawin playing catch with Joe Ainge	225
7	Winter shot with Pearl on the skid driving horses and Golden York standing by.	225
8	Earl Douglass writing data sheets and letters for more money? - 1908	226
9	Jurassic beds with buggy in the foreground - 1908	226
10	1908 - fibula from a dinosaur - fragments pieced back together	227
11	Dr. Holland and Frank Goetschius (Pearl's brother) in the Uinta Formation in 1908	227
12	Discovery vertebrae with Douglass and 11 visitors - popular from the beginning.	228
13	Original Aug 17, 1909 discovery - eight tail vertebrae from a "Brontosaur"	228
14	View eastward from the Green River toward Split Mountain	229
15	Using feathers and wedges to split the rock on the quarry face.	229
16	Caudal vertebrae with pelvis and pubic peduncle in quarry face	230
17	Lowering of femur – three men with a block and tackle on the quarry face	230
18	Pelvis with femur and caudals in place	231
19	Lunch break with watermelon - pick a soft seat, any rock will do	231
20	Earl Douglass and O. A. Peterson about 1920 next to dinosaur wall	232
21	Pearl, Gawin, Earl (mapping) as well as men working on the quarry face (1912)	232
22	Dynamite blast with numerous debris blocks – looking toward Blue Mountain.	233

23	Multi-tasking on the quarry face	233
24	Quarry face with railroad rails	234
25	Block and tackle with tripod	234
26	Two men pushing cart on quarry tracks (only railroad in the Vernal area - ever)	235
27	Rail tracks covered with plaster jackets ready for transport to the Uintah railhead	235
28	Prospecting the quarry face – Earl rappelling down the cliff	236
29	Three men lowering plastered bones with a block and tackle	237
30	Photograph of three men on the track, rolling the crate on a piece of pipe	237
31	Partially plastered pelvis of a large sauropod	238
32	Fencing used to stabilize the plaster on the pelvis of a large sauropod	238
33	Lower the wagons, don't raise the bones	239
34	Photographer capturing a camp scene at the base of the cliffs	239
35	Three men hazarding their lives to find dinosaur bones	240
36	Team of four horses straining to pull a plastered block and two crates of bones	241
37	Wagon train at "Alhandra" ferry crossing – south side of the Green River	242
38	Team and wagon full of bones ready to disembark on the river crossing	242
39	Bones under tarp on the Uintah Railroad from Dragon, Utah to Mack, Colorado	243
40	Freight wagon with bones unloading to the flatbed on the railroad car	243
41	Six teams and wagons with bones headed toward the ferry in winter	244
42	Riding the skid loaded with a block of dinosaur bones	244
43	Bill and Joe plowing the base of the quarry. Notice the mapping grid marks	245

44 Three wagons with bones and a Model-T Ford by Douglass cabin 245
45 Prospecting on *Dolichorhinus* hill, worker lowered off the ledge by rope 246
46 *Dolichorhinus* quarry - 5 large mammals from the Uinta Formation..246
47 Pearl and Gawin in garden in Orchid Draw 247
48 A view of Camp Gulch taken from the hill south of the quarry site..247
49 Orchid Draw – Earl Douglass, Gawin and colleagues 248
50 The Lone Tree (or Powell Tree) on Green River with the Kay cabin...248
51 Pearl pulling Gawin on ice on a home-made sled with runners...249
52 Pearl reading to an older Gawin in the shade................249
53 Snow encroaching on the tent - the early years 250
54 Earl Douglass reading inside their log home - a little civilization in the wilderness 250
55 Earl Douglass all dressed up, standing and contemplating – saying goodbye?...251

Epilogue..417
Appendix...425

INTRODUCTION

Gawin Douglass, my father, spent several years putting this book about his father, Earl Douglass, together using Grandfather's diaries for the majority of the content. However, he gave up trying to find a publisher and left copies of the manuscript with me, my sister Mary Madison, paleontologist Sue Ann Bilbey and at the University of Utah Library. There were other copies (in various states of editing) which I have recently found in files in my basement. Many of the old photographs, taken over one hundred years ago on a Graflex camera, are still in good condition and I discovered my father had spent enormous amounts of time labeling pictures so we can now identify specific people and places. Evan Hall, photographer, took boxes of the photos and scanned them into a computer. However, deciding which photos should be put into the midsection of this book was not always clear and turned out to be a time-consuming process. We think we have made the right selection and trust they will be of interest to the reader. The picture section is in the middle of the book including captions.

Much of my grandfather's writing is either from a scientific or philosophical perspective but in addition my father included many human-interest stories of his own describing life in the early 1900's. I am thankful that he spent so many hours writing and rewriting his manuscript because many of the stories give details only he would have known. I have decided not to rewrite my father's work but rather do some editing on it. The book belongs to Gawin, who has truly honored his father, Earl Douglass through his efforts.

The title of the book is taken from Chapter 35 where my grandfather, approximately 11 years before his death, wrote about his "vision" of a natural museum. In this essay he envisioned entering a large room where he and his colleagues viewed an enormous wall embedded with dinosaur bones in relief. Beside the bones on another part of the wall was an inscription taken from the Bible, part of which read: "speak to the earth and it shall teach thee". These words were quoted from the King James Version of the Bible, the main English translation available in that day, but I chose to use the modern New International Version because I felt an avant-garde person like Earl Douglass would have spoken in the vernacular of the day. In many ways this is what Earl Douglass believed as he wandered the dry, barren, desert areas of Utah with pick in hand. He was sure that with enough time he could find some of the answers to many scientific questions if he could only continue digging away at the earth finding fossils that told their stories.

When formatting the book, in order to show a difference between the writings of Gawin Earl Douglass and his father, Earl Douglass I chose to indent anything that was from Earl's writings or were his quotes. Gawin's writings are the regular print out to the margins. For the most part we, the editors, chose to leave the diaries of Earl Douglass as they were written and did not correct the old spelling or word usage. Where there are three dots, . . . it indicates we either could not understand what was written or it was irrelevant. There are editor's notes throughout which are clearly marked by DDI, for me, Diane Douglass Iverson; SAB, for Sue Ann Bilbey; and a few by GED, Gawin Earl Douglass.
Diane Douglass Iverson, Editor

⌘ ⌘ ⌘

ACKNOWLEDGEMENTS

Looking through the files that sit in my basement and reading the musty papers written by Earl Douglass, his wife, and son, I realize how much work was done in an attempt to preserve and honor Earl's discoveries and writings. I am awed by the hours my grandfather and my father have put in. I have thought for years that I should put forth some effort to try my hand at publishing my father's manuscript about his famous father, but it was my husband, Tom Iverson (PhD, mathematics professor) who finally said we had to get it done in our lifetime. He prodded and pushed then finally began to talk about self-publishing. It was Tom who carefully read the whole manuscript several times through marking items that should be changed and patiently encouraging me when I said there was too much to do. He was the one who helped me to overcome my naturally laid-back temperament and I want to give him much of the credit as an editor and facilitator.

Sue Ann Bilbey (PhD., paleontologist in the Vernal, Utah area) became a friend of my father, Gawin, after she had done an extensive appraisal of the family property which is situated within the boundaries of Dinosaur National Monument. She and photographer, Evan Hall accompanied my father on a hiking excursion into the rugged country of Orchid Draw and listened to stories of his youth growing up in that area. At that time she promised him she would help get his manuscript published. After Dad's death in 1998 she continued to encourage my sister and me to publish the manuscript because of its importance to science, but none of us had the time to put forth much effort. In the past few months Sue Ann has helped proofread, corrected scientific errors, offered advice, sent materials and has been very supportive during this process. The work would not have been as complete or correct without her.

James Evan Hall, a photographer in Vernal, Utah who is also a paleontologist, was invaluable in his care of the Earl Douglass photograph collection. He took his own time to retrieve boxes of old photographs and negatives from our home then painstakingly scanned over 1000 of them into his computer. He then hand-delivered them on a computer hard-drive to us at our ranch in Montana so we could view them and choose which photos would be most interesting and relevant to include in this book. Because of Evan's interest in history and paleontology, his advice and counsel regarding which items should be included were extremely helpful. Evan and Sue Ann, a very hardworking and competent husband-wife paleontological team, have honored Earl Douglass and given him much credit for his work by helping with this endeavor.

Long before any of us began to delve into the files left by Earl Douglass, an archivist at the Carnegie Museum of Natural History took an interest in the Douglass work. Elizabeth (Betty) Hill transcribed some of Grandfather's hand-written field journals that were difficult to read and typed what she could understand so they were legible. When we visited the Carnegie Museum she gave us copies of periods that were not covered in Grandfather's diaries (first to Sue Ann Bilbey and later to my sister and me) so we could fill in the gaps in the discovery events. Betty has been my contact via email from time to time when I wasn't sure with whom to communicate at Carnegie. Her diligent work and care of the Earl Douglass papers are greatly appreciated by our family.

My sister, Mary Douglass Madison, has always been an encouragement to me but she was especially affirming when I called asking for advice during this project. She continued to support us by searching through pictures and letters and advising me as to the title. Mary Margaret (for those in the Vernal area she is named after Mary Ratliff of Vernal and Mary and Harry's daughter Margaret who died at a young age) is very proud of her heritage as the granddaughter of Earl Douglass and at an early age thought she would follow him in becoming a geologist or paleontologist. Instead she learned about the bones of the human body and became an outstanding nurse using her abilities to help people in Washington D. C. and on the Indian reservations of Montana.

The six great grandchildren of Earl Douglass, Paul Iverson, Philip Iverson and Mark Iverson Lueck along with Kathleen Dodson, Carl Dodson and Peggy Dodson Martino have listened to our stories about their great grandfather and have continued to ask how we are doing with the book. They always showed an interest in Earl Douglass when we chose to tell long stories that we had heard in our childhood, or listened intently as we read/recited the poems that are still relevant to our family today. They and their families have been very supportive by showing interest in the completion of this book and remaining positive about their heritage.

Special mention should be given to Dawn Merrick of Vernal who at the direction of Sue Ann Bilbey did the work of transcribing my father's manuscript from his typewritten pages into a computer document making the editing process possible. She saved us hours of our time by getting us into the computer age.

Several friends in our corner of Iowa have been an encouragement to both my husband and me. It seems as though they never tire of hearing our stories about my grandfather and his work and many have brought back reports after visiting Dinosaur National Monument. I have been asked to speak about "The Dinosaur Hunter" or "Geological Man" in literary groups, civic organizations and schools and without this interest I might have tired in trying and perhaps have given up on preserving my grandfather's memory.

Our father, Gawin Douglass donated Grandfather's original handwritten diaries, many journals and writings along with a large quantity of photographs, to the Marriott Library of the University of Utah. Some of the negatives used in this book are housed in the Marriott Library. The Douglass Collection, which occupies 17 feet of shelf space, continues to receive excellent care at the University Library.

Other family photographs were donated to Dinosaur National Monument. Many of these were taken by Earl Douglass but others may have been taken by neighbors and workers. Known photographers of the day were Leo Thorne, John Kay, Jay Kay, J. LeRoy Kay and Pearl Douglass. We want to thank the families of all these people and recognize the work that was done by their ancestors to document and preserve this important discovery. Please note Chapter 34 where my father, Gawin, acknowledges the diligence of the early workers who continued on with their efforts through adverse circumstances.

⌘ ⌘ ⌘

Diane Douglass Iverson
Editor

PREFACE

My father's historic discovery of the dinosaur deposits at Dinosaur National Monument near Jensen, Utah, August 17, 1909 and the extraction of the skeletons was the culmination of a lifetime of personal sacrifice and hardship, but it was not the end of his vision. His dream of a natural museum at the site of the discovery, where all mankind could see the fossils as they were buried millions of years ago, was finally realized when the visitor's center at quarry site was opened to the public June 1, 1958, twenty-seven years after his death.

I remember my father as he walked to and from his work along a dirt road at the base of the high cliff east of our log cabin home, which was located about a half mile from the quarry. As he walked he read a book. His long, loose-jointed legs seemed to throw his feet forward in exceptionally long steps. Few could comfortably keep up with him. He wore an ordinary work shirt, brown corduroy pants, held up by suspenders and stuffed into high-laced, hobnailed, leather boots. He wore an old felt hat and a white handkerchief either tied around his neck or partially tucked under his hat in the back, the rest covering his neck for protection from flies and gnats.

For some fifty years my father's search for truth led him into both literary and scientific fields of investigation; Spenser, Tennyson, and Emerson; botany, geology, and paleontology and finally to teaching, discovery, writing and contemplation. Poverty, poor health, and frustration cast long shadows over his efforts and death ended his hope of publishing his memoirs, poetry and scientific discovery. Although he wrote continuously throughout most of his life nothing has been published except for a few magazine and newspaper articles.

At the time of his death in 1931, my mother and I were deeply in debt, since my father had sacrificed monetary reward in his dedication. I was twenty-two at the time and found myself without a job during a depression. During his illness the previous year mother and I had both worked to make payments on our newly-purchased home in Salt Lake City. Our immediate problem was whether or not to sell it. If we sold we did not know how we could preserve my father's private collection of fossils, his lifetime collection of books, his diaries, letters, notes, photographs and paintings.

Mother took in boarders while I obtained work in eastern Utah. This did not last long however, and we were forced to sell. Had it not been for Mary Sauerbier, a very dear friend and former student of my father who allowed us to store the fossil collection at her place in Salt Lake City, everything might have been lost. There were some 100 boxes weighing between one and two tons. Miss Sauerbier kept them in her basement for close to ten years guarding them carefully without pay.

When the Utah Field House of Natural History at Vernal, Utah was completed, Mr. G. E. Untermann, Director of the Field House and an old friend of my father, graciously offered to move the material there for storage and exhibit. Except for the diaries, notes, pictures, and books the collection still remains there.

My mother and I moved most of my father's records to the ranch home of her brother, Grover Goetschius, near Virginia City, Montana. They remained there until 1960 when I moved them to my home in California to begin my task, the completion of the book my father planned to write. My mother also longed to complete the task but her health failed. She died in 1955 but her touch remains in the diaries.

To support my mother, and later my own family, I worked as a mining engineer in the western United States and South America. It has only been quite recently that I have had the time to study my father's records and write. Although I had been familiar with his diaries for years, his hieroglyphics discouraged me from making a real effort to transcribe them. To my surprise, however, I found when I became acquainted with some of the peculiar characters that kept reappearing, transcribing was not as difficult as I had supposed. In fact, I found pleasure in puzzling over and finally discovering the meaning of a word or phrase, probably in much the same way he felt when making a scientific discovery.

Rumors and half-truths have, at times, partially buried my father's success. Much of the credit for his work had been taken by his contemporaries and assistant. His discovery of the famous fossil beds at Dinosaur National Monument was even attributed to Elder George A.

Goodrich, a fine old man who assisted him. Arthur Goggeshall, a fellow worker, wrote a misleading account which was published in the National Parks Magazine, Vol. 27, No. 113, April-June, 1953, p. 59. When I visited the Carnegie Museum during the summer of 1967 my father's name did not appear with his remarkable paleontological discoveries on exhibit there.

This book is in memory of my greatest teacher, a tribute to a dedicated scientist, whose thoughts and records should become a part of our country's heritage. It is the odyssey of one man's life as reflected in his diaries begun when he was twenty-one. It is the tale of a young man reared in a religious home and tormented by the implications of evolution. It is the story of a dedicated fossil hunter, his struggle to know the truth, his happiness and sorrows, his marriage to Pearl Goetschius, a former student. It is a record of his teaching experience, his years at Princeton, and his struggle to satisfy the needs of the Carnegie Museum, the National Museum, and the University of Utah.

Gawin Earl (Doug) Douglass, 1908-1998
(only child of Earl Douglass)

⌘ ⌘ ⌘

EARL DOUGLASS:
A Heritage of Tenacity and Love of Freedom

For forty-seven years my father, Earl Douglass, kept a diary; a daily, weekly, and monthly record of a man devoted to many aspects of science, fossil digging in particular. To him fossils represented the secrets of life hidden by millions of years. Reared a Seventh Day Adventist, the implications of evolution tormented him. Hence, his determination to devote his life to a search for truth.

Perhaps it will be easier to understand my father if the influences that shaped his life are reviewed. Although the Douglass history began in Scotland, my great grandfather, Alexander Douglass, grew to manhood in the New World. It was his heritage of courage, love for freedom, and tenacity that set the pace for his sons and his son's son.

Grandfather Fernando Douglass, who lived with us during the last years of his life, told the story of his father, Alexander, who resisted the armed might of Great Britain. It was during the War of 1812 that England decided they needed Alexander, at the time living in Canada, to fight in the armed services against the United States. He responded with, "I refuse to fight against a good government like that of the United States."

"Great Britain can handle boys like you," the official answered.

"There aren't enough men in Great Britain to make me," Alexander retorted.

With that his course was set. When darkness could hide his movements, he rowed across the St. Lawrence River in a skiff and as soon as possible enlisted in the U.S. Army to serve until the end of the war.

Grandfather Fernando Douglass spoke of his father with pride as he told of his deeds of courage and determination, even to the extent of stubbornness. He could go for years without speaking to someone who had done him an injury but at the same time his heart filled with compassion if his worst enemy was ill or needed help. During his lifetime Alexander became a teacher and a popular doctor, bleeding more men and pulling more teeth than most of his contemporaries. Together, he and his wife Laura Stanard, also a teacher, reared eleven children, six boys: Leander, Earl, Orlando, Fernando (my grandfather), Evander and Ozander, and five girls: Candis, Appoloni, Aphia, Felecia, and Laura. When asked why he chose such long names, Great Grandfather answered, "When someone calls the pigs, I don't want the kids to come running."

Grandfather Fernando, the sixth child, settled on a farm near Medford, Minnesota, where he had taken a preemption on wooded land on Rush Creek. Here basswood, elm, maple, box elder, oak, iron wood, hickory, hackberry, white ash, thorn apple, plum, and black haws grew in abundance. He paid about $200 for 160 acres.

Clearing the land and hauling logs by oxen to the mill was the primary occupation of the early settlers in this area. In 1857 Fernando married Abigail Carpenter and began his family. To them were born three children, two girls: Nettie and Ida, and one son, my father, Earl Douglass. The family worked and played together remaining deeply concerned for one another all their lives. One of the cementing factors was the belief in honesty and truth. If a person professed his belief in a certain church or religious group, he must practice his religion and not be a skeptic. To Grandfather the Bible was more important than tradition.

After going to Minnesota and attending a course of lectures by a Seventh Day Adventist minister, Grandfather joined their church, accepting the faith and actively participating in its activities for the remainder of his life. For this reason he did not enlist in the War of Secession in 1861 for he felt its only purpose was to preserve the Union, not to free the slaves. He became a strong abolitionist until the slaves were freed. He was a devout Republican until that party refused to take a stand on prohibition, then voted or supported no party, for in his heart he remained a prohibitionist.

Grandfather Fernando spent the latter years of his life with our family in Utah. After his death, The Daily Peoples Press, Owatonna, Minnesota, March 28, 1916, reported:

> "He was too generous, sympathetic and brotherly to accumulate much wealth. No one in sickness or trouble ever sought his aid in vain, if it were in his power to help them, and he would uncomplainingly sacrifice his own comfort and convenience for those who were in trouble. He hated sham and hypocrisy, but admired genuine honesty and sincerity. He respected an honest sinner more than a pious fraud."

Grandfather was buried on our property in Utah, which is now a part of Dinosaur National Monument. In his later years, Father often spoke of his family and boyhood days in Minnesota. He wrote of the beauty of the land and the fun he and his two older sisters, Ida and Nettie, enjoyed together:

> "Medford was a little town of three or four hundred inhabitants. A strip of beautiful forest three or four miles east, containing scores of species and varieties of trees and shrubs, was a land of poetry and romance to a boy. Westward was the great prairie or rather a land of oak openings, aspen groves, swamps, and meadows. Hundreds of beautiful wild flowers bloomed in the green meadows. Some of these trees and flowers now seem sacred, the groves of quaking aspens, the wild crab apple with its pink flowers making a beautiful cloud of bloom, the wild prairie lily, and others. The farm on which I spent my days until about twelve years of age was sacred to me, and long after we left it, I planned and longed to return and live in the land of childhood dreams. Perhaps I was thirty-five or forty before this longing left me... We used to wade in the pond. I learned of the plants and animals that inhabited the water or lived around it . . . I was born among the hills, but I've always wanted to get up higher. I climbed all the highest hills, but wanted to go still higher, to touch the clouds, to get above them."

My father's formal education began in an old board shack when he was five, but his thoughts danced with the fairies in the groves while he pretended to listen to Mrs. Farnum, his teacher. Writing about education, after he himself had become a teacher, he said:

> "In those days the first thing we must do was to learn the letters of the alphabet. Of course, letters were the foundation of all knowledge, at least book knowledge, and all knowledge or wisdom was supposed to be in books. So we began at the foundation. The first thing we learned was to sit still and do nothing. We had to rid our young minds of all ideas and our young and vigorous muscles of all activity until we learned those conventional hieroglyphics that would in the future years convey to us all we would need to know of earth, heaven, and hades."

His second year of school was somewhat more pleasant. Father remembered his young teacher holding him on her lap and pointing to the letters. A significant part of his education, however, was on the farm itself where he learned to cut the grain with a cradle, raking and binding it by hand. Perhaps he learned the importance of industry while working with his family. At one point, his book education caught up with his dreams, for he wrote:

> "Two other boys and I built a one-room cabin close to a thicket of jack oak and poplars. I chopped cord wood in the heavy timber a couple miles from our cabin. I got 60 cents for soft wood and 80 cents for hard wood, mostly oak. Some mornings every branch of the forest and every dried bush and herb was decorated with frosty crystals, sparkling like gems in the sunlight. It was now, that I thought of Spenser's Fairy Queen, when I walked and worked in a fairy world. I remember the lines I loved:
>
>> There is a pleasure in the pathless woods,
>> There is a rapture on the lonely shore,
>> There is a society, where none intrudes.
>
> As long as I can remember poetry was a source of delight. It was the expression of the beauties and charms of nature, a world of unexpressed meaning. Poets were the great inspired men of earth, the great geniuses."

Father's scientific education was reawakened in the two-roomed school in Medford where he thought of his teacher as his Fairy Queen.

> "She wove a hallowed atmosphere of romance through my life which was potent for many years... I borrowed her Dana's Geological Story Briefly Told, and sat up until three or four o'clock in the morning to read it through so that I could return it the next day, for it was her college textbook."

He had been excited by geology at a much earlier age for he often spoke about going with his father to Clinton Falls where a dam had been built to develop power for a grist mill. Nearby was Lindersmith's limestone quarry, where quarrymen pried up layers of limestone and broke them into blocks for building purposes. He wrote of his visit to the quarry:

> "Mr. Lindersmith told us about the long slabs of limestone, flat on top and convex below. He said that geologists said they (the fossils) were animals which lived in the sea years ago. He thought they were some kind of "saurian." I had heard of Ichthyosaurus (fish lizards) and Pleisaurus (near lizards). Years later I found out that these "lizards," and "near lizards" were giant mollusks, distantly related to the present Pearly Nautilus, but with straight rather than coiled shells-hence the name "orthos" (straight) and "cress" (a horn). This wonder and mystery of animals that had lived in the sea long ago aroused my imagination and kept it aroused."

Father's fancy turned naturally to gathering bones he found on the prairie around his home:

> "We used to wade in the ponds. I loved the ponds and marshes, streams and lakes, in fact almost any place around water. Trying to restore ancient scenes and landscapes fascinated me. On the pasture hills, I used to gather whitened bones of sheep and play with them.... Sometime after I was fifteen we moved back to the little town. I went fishing as I often did. The fishing was not good, but when going around

a muddy pond I saw something shiny and picked it up, for it was especially pretty. It proved to be a variously-colored, banded agate. In excited wonder, I took it home. It became the first specimen in my permanent collection."

My father's interest in collecting specimens of scientific interest never faltered during his entire life. From the moment of his discovery of fossils in the stone quarry he knew no respite in his search for truth recorded in the rocks and fossils of the deserts and hills. His search continued in books, for he knew that men had already discovered much. He went to college and became a teacher, primarily as a means of earning a living and because of his fervent need to do research, to continue his discoveries in the field.

⌘ ⌘ ⌘

A COUNTRY SCHOOL TEACHER:
Facing His Dilemmas

According to my father's records, he earned his teaching certificate when he was twenty. He had passed the state examination but was "conditioned" in grammar. In later years he often spoke of his early education:

> "I liked most of my teachers–loved them I guess– but it has taken all my life to get rid of the false principles and ideas, to overcome the dullness with which false methods of education clouded my mind–and I have not gotten over it yet, but I was learning something from outside, from the great teacher whose school was around me... Yet when all is summed up, I did learn to read and with that all the knowledge, lies, false theories, and bunk that was in books was open to me."

My father's first teaching position was in the Swiss-German district of Deerfield, Minnesota, about six miles west of Medford. The following spring, 1883, he began teaching in the Richland district, eleven miles northeast of Medford. The following year he began keeping a diary. Many of the entries in his first book began with a comment about the weather which was followed by a verse or quotation. He explained his purpose for keeping a diary in the first two entries.

> Tuesday, January 1, 1884 - Cloudy and cold, not much snow yet.
> "Let the dead past bury its dead." Longfellow
> "Ring out the old, ring in the new." Tennyson
> This is the first day of the new year. The old is gone forever. Its record is unchangeable. I have not acted my part

nobly in this great drama of life. I can hardly point to one noble deed or unselfish act. My motives in general, I fear, have not been right. I think after all it has been the most important part of my life. It has been one-sided and has created serious doubts and fears. Doubts have arisen in my mind concerning the Bible and fears as to the destiny of man. By earnestly praying to God and studying His Word more of these doubts are being removed. I mean to make a point of studying the Bible more this year.

Wednesday, January 2, 1884 - Clear and cold, wind in the west; frosty.

Our lives are books. Each day a page is written, good or evil. This book purchased at Minneapolis, December 25, 1883, price 85¢. Perhaps it will not be kept as diaries commonly are but I shall try to use it in the way that will be most beneficial. I want to record here the things that I need to remember or want to refer to. If I have learned anything through the day I want to write it down to strengthen my memory. Thoughts and suggestions, plans and references to my work, articles written with view of having them published may be written here. Criticisms, short poems, and stanzas of longer ones, quotations, thoughts on religious subjects, etc. In fact anything I choose and consider beneficial.

I think we have taken an advanced step in the new year. Still I think there are some things wrong in the government of the school. I must set about with writing, vigilance, and good judgment to set things right. I feel weak and incompetent for the task.

At this time in my father's life he was torn between the strict orthodox religious atmosphere in which he had been reared and the scientific truth he had discovered relative to the theory of evolution. He wanted desperately to reconcile his religious beliefs with scientific facts but this wasn't easy for a Seventh Day Adventist. Each day he faced this dilemma in the classroom: What is the truth? In addition to this he had the same problems facing every beginning teacher who wants to succeed. He had set high standards for himself and thought that he was not achieving them.

> Monday, January 7, 1884 - Weather milder.
> "But whatsoever may keep tomorrow know I not."
> School–How my work today should be criticized at Normalville. The teacher was not animated enough, occupied too much time for some recitations, was not well enough prepared for work. The B geography had a much better lesson than common on account of making questions and asking each at a time ... and having them arise to answer it. The C arithmetic have had their lessons good. Is it not possible to bring all the classes to the same degree of perfection? Improve!!! Improve!!! Improve!!! It needs energy, genius, skill, thought, knowledge, and success. I wish on every page of this book I could write improvement.
> "Not enjoyment and not sorrow
> Is our destined end and way
> But to act that each tomorrow
> find us further than today."
> Henry Wadsworth Longfellow

Friday, January 11, 1884 "Health is wealth."
A blizzard has been raging this afternoon. Last Wednesday I sent a few words to my father stating that I would be home tonight if the weather permitted. I fear that he started after me. I did not deem it prudent to start out with the roads drifted full and the cold wind blowing a gale and my lungs stopped up with asthma. I was not in good condition to travel through the deep snow so far.

I have been making somewhat of an effort of late to regain my lost health, but I have not been strict enough. I fear my good health has gone forever and through the remainder of my life to be a broken down, complaining, unhappy, useless being living only to be a hindrance to others. I think my life must be brief. My stomach is out of order. My hand trembles like that of an old man and my heart beats wildly. If man does not live as he should he must suffer the consequences.

Sunday, January 13, 1884 - Warm, thawing nearly all day; 40° above.

Read until about 11:00 A.M. then walked over to the Norwegian store, bought two handkerchiefs at 10¢ each, 15 two cent stamps and returned by Mr. Beards; stopped about half an hour. The girls and George were not at home . . .

Monday, January 14, 1884

Received my mail tonight by Mr. Dalton, 12 writing spellers, 1 Good Literature, 1 Teachers Inst., 1 Peoples Educator, 1 St. Cloud Normal School Catalogue. The School was far from satisfactory today. Felt discouraged and down-hearted. If I do feel so, I must not show it to the school.

Tuesday, January 15, 1884 - Quite cold
"Think before you speak."

Blunder, blunder, blunder. When will I cease blundering. It is killing my school. It is making a fool of me and lowering me in the estimation of my pupils. I must think twice before I speak once. To guard against mistakes I must look over the lessons before having recitations. Used the spelling blanks today. I think they will be a success. The writing was done nicely. There is an improvement in story writing. I am afraid the scholars are not studious as they were and more mischievous. I must set earnestly at work to correct these errors. Perhaps if I can induce the parents to come in occasionally it will be a means of improving the school. I am nearly convinced I am not a good teacher and never will be.

Tuesday, January 29, 1884
"Life is earnest."

Went down to Eastman's in the evening with horses and cutter and stayed all night. Laughed and joked as I always do when I am there. I am too light and trifling. If I had a burden for the souls of my fellow men that I should have I would not be so trifling.

Saturday, February 2, 1884
Went to meeting in Medford. Did not feel very loving and Christian-like. I have no friends and there is doubtless reason for it. . . . If the selfish thoughts and feeling of today should remain in the Great Account Book of heaven they should fix my doom for all eternity. These are solemn days. Time is fleeting, moments hasten, days are passing, years go by. Think a moment, stop, and listen. Think how swiftly the moments fly. Youth is short and age is dreary; cheer these early days with a song or thou wilt grow sad and weary if the journey should be long. Up and arm the fort, the battle for the strife with you.

Monday, February 11, 1884
It has been storming some this afternoon. There is little or no teaming on the roads now. Not very cold. Have been reading from Lyell's Geology. His theory is antagonistic to the doctrine of a revolution in the earth's structure. It seems to me more and more that this doctrine has exceptions at any rate. How can fossils in such great numbers be deposited in the earth and be preserved except by catastrophe? How can mammoths which are found frozen in the earth get there without the earth thawing and if it was subject to thaws it would certainly be decomposed.

The school term was completed March 11th, 1884. My father wrote that he was inexpressibly sad and lonely the day school closed. On March 20th he received his school "order" (warrant or paycheck) $125.00 for the term. After paying most of the bills he had accumulated he agreed, at the suggestion of Brother Olson, to go to Rochester and canvass for the Church. He and a Brother Hilliard went to Chatfield, south of Rochester, for about a week to sell a church paper. They had little success. The rest of the summer was spent in reading and studying together, planting a garden, painting the church, and making some improvements around his folk's home. Later in the summer and fall he and his father harvested grain for themselves and for various people in the neighborhood. When my father wasn't working he made frequent trips to the river, sometimes to hunt or fish. More often he went to examine the few rocks exposed in the river beds and to study the processes of erosion and deposition.

Father was elected secretary of the Sabbath School. He did not especially enjoy this position for he dreaded reading the reports. In addition, he was continually getting himself into trouble by expressing his views regarding science and religion. He probably found more enjoyment at the river, which was full of minnows. It was at this period in his life that both evolution and the moral foundation of society began to disturb him.

> Thursday June 5, 1884
>
> Went over to Fred's store and sat a long time reading the newspapers. It seems that the moral foundation on which society rests is like loose, unstable sand. Men are losing their morality as fast as they are gaining in the arts, invention, and science. Freedom and patriotism are dead. Love of money and pleasure are the craze of the age. No fit men can be placed at the head of the nation. Virtuous men and chaste women are growing fewer and fewer. The only thing that can prevent this, it seems to me, is faith in the Bible. People are passing from superstition into infidelity. Men always go from one extreme to the other. They judge the Bible by their grandfathers' opinions of it. It may take even greater knowledge of science than man yet has to understand its teachings. We cannot judge the Bible by present opinions relating to science, for they have not been fully established.
>
> Discovering the truth assumed greater significance each day. He continued his search for fossils and read widely. It was later in June that he found his first fossil which remained a permanent part of his personal collection. Near the river

he found a stone full of shells, apparently of a species not inhabiting the river. When it rained, he read.

Thursday, June 19, 1884
Have been reading a good share of the time. I am reading the life of Josephine. She is my ideal of a woman. I love her because she was pure, noble, kind-hearted, loving, gentle, intelligent, and sensible.

Continuing his education became a major problem. My father still owed money to his last boarding place and he finally decided that he had to leave home if he were to get ahead. After harvest he paid his last board bill and registered at the academy in Owatonna on September 2, 1884. He finished October 24th. Father had sold his prize colt for $80 to pay for expenses. In November he resumed teaching at Richland but returned home on weekends, weather permitting. Although he believed that his scientific interest had lowered him in the eyes of the church, he was elected unanimously to be the Superintendent of the Sabbath School.

Friday, January 2, 1885 - Weather still, -42°.
Last night was probably the coldest of the season. I bought 1 pair of felt boots, 1 pair of rubbers, 1 pair of slippers, 2 pair of socks. Went to meeting at the Church this evening. Had a very good meeting although there were but few present account of the severity of the weather. I wish I knew whether or not the Church has the truth. If they have in every respect I fear for my own eternal welfare. It may be my fault that I doubt so much, but how can I believe against strong evidence? For instance, how can I believe the earth was created in six, 24-hour days?

Sunday, January 25, 1885
Came to Williamses; went through the woods and had to inquire the way. Got lost on the wrong road once. Got stuck

in the snowdrifts by Methodist Church, tipped the cutter over twice and got the horse down. Had to lead him out and pull the cutter and leave the things, leave the sacks of corn at house nearby. Froze both ears. The wind blew from the N.W. and was sharp.

Thursday, January 29, 1885
   School more satisfactory. We are reading "Voyage of the Sunbeam" at Williamses and I am reading "Town Geology" by Chas. Kingsley.

Friday, February 13, 1885
   Have not been feeling so well today and have not enjoyed the school very much. Have tried to be patient and do my duty cheerfully. I tell you there are battles to be fought in the school room. I hope it will be easier to be patient next time. I have naturally a terrible mean disposition.

Sunday, February 22, 1885 - A pleasant day and warmer
   I found, to my sorrow, while looking over my accounts that I had made a fearful mistake which will probably prevent my attending Normal. I thought I would have $60 but will have not more than $20 or $40. Don't know what to do.

School ended March 17 and somehow Father managed to attend the academy. He registered in political economy, civil government, algebra, physical geography and rhetoric. Thoughts of evolution, religion, and morality continued to plague him but he felt an increasing need to observe the world around him more closely.

Monday, May 25, 1885
   Today the earth is a paradise. Almost every disagreeable thing is covered with Eden-like beauty. Woods form rolling, waving masses, reaching up into the transparent air against the clear heavens and going down into the valleys; the river winds through the forest of living green. There is the deep

green of the poplars and the transparency of some other trees. What a wonder, the bursting forth of unnumbered myriads of leaves and flowers. The plum and apple trees are in their bridal robes and the air is laden with sweet odors.

Saturday, May 30, 1885
    I said some things this morning that were useless; I must stop speaking of those subjects to those who do not love to hear them–science, evolution . . .

Friday, June 26, 1885
    Asa Armstrong was here this evening. It was the first time I had seen him for three years. He looked over part of my collection and seemed quite interested in it.

Father spent the remainder of the summer doing odd jobs, harvesting and haying. However, he did attend the Sabbath School Association meeting in Fairbault and the camp Meeting in Mankato. He was again elected Sabbath School Superintendent but this time there were four dissenting votes. Once again he agreed to teach the Richland School.

⌘ ⌘ ⌘

3

EARLY SCIENTIFIC INTEREST:
A Love Affair

When the fall term of school opened at Richland my father's scientific interest was reflected in the classroom but his emotional interest turned to affairs of his heart.

Saturday, October 9, 1885
We had probably one of the best Sunday Schools we have had. Had a lesson in Charity and it seemed to do good and nearly all seemed impressed.
Had a sing in the evening and enjoyed it very much.

Wednesday, October 21, 1885
Went to Fairbault and got a new certificate. The Supt. examined me in but one study, Grammar. I raised the average from 80 to 88. Went to Dr. Cook's. He showed me his microscope also objects viz. fly's eye, bean starch, wheat starch, etc.

Wednesday, October 28, 1885
School improving.
Today is my 23 birthday. It makes me feel sad. 23 years of my life gone and what use. Oh, may God show us what life means.

Wednesday, November 11, 1885
Went up to the store and got two registered letters made out.
Sent for Slete's Hygiene Physiology. Took two copies of the Abridged Edition at 50¢. Sent also to H. H. Fanman for 20

Rocky Mountain specimens. I cannot express my anxiety for the arrival of the specimens. I have long desired some and now I hope to have my desires gratified. It has made me happy in anticipation.

Wednesday, November 18, 1885
   Had a little dissection in Phys. class. A hog's lungs and heart, although imperfect, were exhibited and talked about. Felt the need of a microscope.

Friday, November 20, 1885
   Went over to the Norwegian Store after school. Sent for:

a microscope	$4.00
12 slides	.25
12 glass covers	.20
1 pr. brass forceps	.25
Canned Balsanac Tube	.25
Postage	.60
Total	$5.55

This is a humble beginning in microscopy but I intend to get a higher priced instrument if I get so I can afford it. I have longed and wished and hoped for a microscope for years and dreamed of the pleasure and instruction it would afford. As long ago as when I worked for David Armstrong I thought when he paid me some money I would get a microscope and would sit down on the harrow and would think of the things I would be able to view through it.

Sunday, November 22, 1885
   Went up to the depot in the morning to see if my specimens had come. The depot was locked. I went over to Mr. Barrow's and Burton said there was a package for me and he would be over soon. My joy was almost unspeakable. Such wild expectation and enthusiasm, restless longing to see the wonders of the Rocky Mountains. I poured over the pictures of them until I had almost worn out the paper (Youth's Companion). I knew that no two rocks could be precisely alike

but my disappointment can hardly be told when on opening them I found they were a lot of little dull looking stones being no resemblance to the illustrations whatever.... I think they were worth the money but not worth the disappointment.

Thursday, November 26, 1885   Thanksgiving
Cora Beard, Ettie Williamson, Mark Beard and I went to Kenyon. I had never seen the place before. There is a large bridge there. I am seldom happier than I have been the last few days.

I went to a temperance social or whatever it is. Did not stay until it was out. Did not like the way of taking supper. Each basket had a girl's or old woman's name on it and the one who bought it ate supper with her.

Wednesday, December 2, 1885
Cora, Mark and I went down to Will and Irwin Mather's.... Mark drove the colts. I seldom have enjoyed myself so well as the last two or three weeks. To teach a good school through the day with little worry and fear, to spend the evening in pleasant company (I won't say who) and sleep sweetly at night are great blessings.

Friday, December 4, 1885
Taught school in the forenoon. Let out school in the afternoon on account of Mrs. Atherton's funeral.
Went to Post Office and got my microscope.

Saturday, December 5, 1885
At Mr. Beard's. Looked in microscope and read.
As this is intended for no eyes but mine and it is too precious to be lost, I wish to record a little of my present feeling, which is kind of an epoch in my life. I have loved more or less many times an ardent but transient affection that perhaps aroused something holier and nobler in my nature. But each has been somewhat different.... The earth seems brighter

and my life for sometime has been happier. It is the sweet pleasure of a calm love which seems to be growing deeper. She who has such an impression on my heart is not perfect, although every art now passes into a different purer measure and is transformed into beauty. But there are certain things about her without which I could not, I think, love her as I do. She is no light fairy thing, all vanity and show, but a good substantial, sensible, young lady who does not follow every whim of fashion and folly of pride. I love to see her with her dish apron on, or a rich yet sensibly made dress. In one she seems useful in the other, to me, beautiful. But how long this will last I do not know. I know one thing, that after so much pleasure there seems to be a cloud coming which I dread. One thought brings a pang to my heart every time it comes to me. She is soon to go away and I shall see her as I have seen her no more . . . I cannot tell, she might be mine. Notwithstanding, my vow is always to remain as I am and my resolve is never to be tied by that hard knot of wedlock, yet the heart that loves loses all resolve but to obtain that object of love. Yet it will probably pass off like the aura at dawn and the heart be freer but sadder. But with her, I think I should love ever. But if love does not cease circumstances, hard circumstances, may separate us.

    I never ought to have thought of this but I could not keep it back entirely. Our belief is different. I am poor and can hardly, with the educational apparatus I need, support myself. Of course I cannot tell what the years may bring, life is a scene of change.

Saturday, December 12, 1885
    Enjoying myself as well as usual . . . Am passing through a new stage of existence.
    Mr. Eastman is digging a well. Went over there and they showed me wood that was gotten from it. I went down in the bucket and saw the wonder. One log was about 35 ft.

below the surface. It was 6" in diameter. I got several pieces and another piece 2 or 3 feet lower.

Sunday, December 13, 1885
    At Mr. Beard's. We had a quiet dispute on the marriage question in the evening. In the argument I was the only one on my side. I contended there were few happy marriages and tried to present the evils and sorrows of marriage. Mrs. Barlow said she and her husband had lived together five years and neither had spoken a cross word to each other.

Tuesday, December 15, 1885
    There are changes in love and it is hard to tell how a fellow will feel next when he is in the dilemma. To tell the truth, I have never thought or felt as I have of late... I sometimes think I would be glad to unite with her whom I love and then I try to banish the idea and settle down as I was before to books, science and the pleasures of the imagination... Then I see her and she seems so good, so kind, so near what seems desirable in a woman and I think of the unloved bachelor, and the man with an intelligent, loving and helpful companion and compare them and feel different...

The State Teachers' Association meeting, in St. Paul was attended December 29th, 30th and 31st. A nice gift was purchased for Cora Beard, the young lady so fondly mentioned, but nothing more appears in the diaries concerning this love affair.

⌘ ⌘ ⌘

HOMETOWN TEACHER:
Humiliation

An interruption in the diary that started soon after the first of the year 1886, continued until March when the school term at Richland was ended March 5th. Father's diary relates his return to Medford to apply for the school there. His fascination with the microscope continued.

Monday, March 22, 1886
Had some more accessories for my microscope made. Went over to Mr. Bailey's and had him make a table for heating slides and bought a little oil for a spirit lamp. The table and lamp cost 26 cents. If I had got them from an optician they would have cost $1.50 at least. Also made 7 dissecting needles.

Thursday, March 25, 1886
Went to Mr. Skinners as he sent for me. He and Mr. Webb wanted to secure me to teach the school. We found there had been a misunderstanding. Mr. Skinner understood me to say that I did not expect to get more for teaching the school than they gave last summer or $32. I remember making no such statement... wanted $35.

Thursday, April 1, 1886
Saw Mr. Bryant. He said he talked with Mr. Webb and compromised the matter allowing me $33.50 instead of $32 or $35.

Friday, April 20, 1886
   Thought I would go up by the old dam and see if there were any fish and thought I might possibly find a frog as I have been anxious to watch the circulation of the blood. . .While going along the pond I saw a frog swim clumsily into the deep water and stop. I stood motionless a long time for him to come to shore. . .After awhile he got quite near but not near enough to reach him from the shore. I rolled a stone carefully toward him, stepped on it and grabbed him in my hand. I prepared a bag and made a slide and watched with great interest the circulation. It was truly an amusing and instructive study.

Monday, May 3, 1886
   School began today with a larger attendance than I had anticipated; got very tired. I see it will be no small job to manage the school but there is no turning back in this warfare. The scholars are bright and intelligent which is much. I think all will be well.

The advantages of teaching close to home were viewed with pleasure but the disadvantages of teaching where he was so well-known and where personal quarrels and jealousies might be involved were not anticipated. The students, being quite noisy and unruly, made stern disciplinary measures necessary. Discipline was apparently not one of Father's major assets as a teacher.

Tuesday, May 11, 1886
   Mrs. Bryant was over here and said she wanted to put a flea in my ear. She told me she thought the school was good only too noisy. She is interested in my having the school and wants me to straighten the children out.

Friday, May 14, 1886
   Kept several children after school as I told them I would punish them if they disturbed those in back of them.

Saturday, May 15, 1886
   Went to S.S. and meeting.
Got to thinking about my school and the enemies I will probably get by teaching and thought of the hard lot of the teacher and felt to almost despair and give up teaching and earn my living some other way.

Sunday, June 20, 1886
   I went over to Mr. Bryant's to borrow his fish spear and he asked what I would teach the lower school for, for one year. I did not know what to say... I was almost shocked. Then he told about another man applying for it. He said it was not that they thought I was not competent to teach the higher grades but that they wanted me for the lower. This was the beginning of sorrows. We talked for some time. At last he asked if I would teach it a year for $40. I gave him no definite answer.

Monday, June 21, 1886
   I never before experienced such a day as this. Since yesterday I had somewhat expected they would hire someone else to teach the room where I am teaching but it hardly seemed they would either. I was somewhat prepared for the shock but did not realize what a terrible one it would be either. I do not think I ever experienced such a complete blasting of all my dearest hopes and fond expectations. My reputation as a school teacher, lost. Had just got to where I felt more at home (in the classroom); where the pupils could be more easily guided on the road to knowledge; where I enjoyed teaching them. Can I ever leave home and go out into the world again? The thought sickens me. Here I am, in my own little village at home all the time, no long and tedious journeys back and forth. Nowhere near the trouble of going out in the country. I had thought if I should get the school I should have apparatus to study the sciences, and above all get a first-class microscope for which I have so long wished.

But it's all gone now. I thought I was rising, but this is an awful fall. Oh, how I wish I had never been obliged to apply for this school and how I regret not keeping better order at all hazards. This is the bitterest thing I have known.

Tuesday, June 22, 1886

Did not sleep good until late the night before and was not fit to teach; was almost sick. Tried to do the best I could. That deep hard sorrow was somewhat lessened by meeting the children. They were as kind and as good as ever.

Wednesday, June 23, 1886

Felt some better. Began examinations. Heard some things that made me feel better. Found that many were surprised at what was done and were opposed to it. This gave me more courage. Mr. Freeman said Fred was mad about it. Mr. Freeman said if they got a teacher everybody liked they would keep him but one term. Mr. Beeman thought it mean.

Thursday, July 1, 1886

I went to Fairibault with Pa. Went to Mr. Holmes to get my hair cut. He spoke about reading in the paper about my school being out and my successor being chosen. I took the paper and read it. It spoke also of the school being a success and of the circumstances.

⌘ ⌘ ⌘

A STEP WESTWARD:
The Prairies of South Dakota

After the bitter experience of losing the coveted school in his hometown for the ensuing term, and the humiliation of not finding another suitable position in the few country schools in the surrounding area, my father left for South Dakota. His oldest sister, Ida, and her husband, Alfred Battin, resided on a homestead near Iroquois. There had been a minor migration of his acquaintances in Medford and vicinity to South Dakota and these friends, together with Ida, had written at times suggesting that he visit them to see what he thought of the country. He had been dreaming for some time of the West and the opportunities it afforded in the fields of geology and paleontology however, South Dakota was not exactly the place he wanted to visit. He did not go immediately, for he was troubled and filled with indecision. After doing his usual summer work in the hay and grain fields, Father continued his adventures with the microscope. On July 4th he made his first entomological expedition, collecting a number of butterflies and dragonflies and preparing them as specimens for study. It was not until the latter part of July or early August when he apparently made his decision to go to South Dakota.

> Thursday, August 12, 1886
> At home preparing to leave. Pa and Ma and I went to Owatonna, attended the meeting and they went home. I stayed at the depot until train time. Did not get much sleep. Bought ticket to Iroquois for $7.75, second class but it was as good a place as any.

Friday, August 13, 1886

I started for the West at about 1:30 P.M. Got to Iroquois, South Dakota at about 12:03 P.M. Gilbert was in town and came to the hotel while I was eating dinner. Coming out I met Ida at Norman Gibson's.

Saturday, August 14

Went to S.S. and meeting in school house where I saw a good many of my old acquaintances.

Sunday, September 19, 1886

Ida and I went over to her claim and stayed overnight. I shot a prairie chicken and we had a good dinner. Enjoyed it very much.

Friday, October 28, 1886

This day I was 24 years old. Oh, how I wish it was not so. It makes my heart ache when I think of it. What has been my life and where has it gone? Where is that greatness and nobility I have been dreaming of? Where are the wonderful things I was going to do? Alas, is not life only dreaming and death waking up?

Sunday, October 31, 1886

Bought a new breech-loading shot gun for $6.00, including loading apparatus, from Mr. Warner.

Monday, November 1, 1886

Took my new purchased gun and went hunting. Went over to Mr. Warner's and got the apparatus belonging to the gun, reamer, loading tools, etc., before I started hunting. Went down to the lake south of here, got in a boat and went in among the weeds and waited for geese. Soon a flock came flying toward me. I was reading "Sesame and Lilies." I laid my book down and got ready to shoot. They came nearer but

I think they saw me and turned. I fired and one dropped. I think I was getting ready to fire again and the first thing I knew the boat was tipping and I found myself on my back in the water. I waded around for awhile but did not find the goose. I was pretty well soaked and my ammunition was wet. Going home I saw a coyote. I shot and it fell over and barked.

Friday, December 31, 1886

Another year almost ended. . .How little I knew where I would be in one year. How little I knew the things that would transpire between then and now. . . I loved and am not ashamed of it. I now look upon it as the happiest time of my life. I cannot say that love is all gone although the object of my affections is far away and a long year has gone since I beheld her. I do not know but I sometimes think she would make a loving and faithful wife but circumstances have separated us.

I do not feel that the last year of my life has been one that I can look back on without shame or sorrow. The future I cannot see but I hope it will be better.

Saturday, February 5, 1887

Was late to Sabbath School. The school board voted me another month's wages. I wanted to go to Iroquois to get this diary as I thought it must be there. I hitched up Wild Bill to the buggy and went down after sundown... Bought Wood's "Insects at Home". Think it is an excellent book and one that I have long wanted.

February 16, 1887   The Mirage

The mirage this morning was the most remarkable I have witnessed. I watched it for some time with great interest. Homes and other objects appeared "drawn up", being narrow and high; again they would be rolling and low...

Monday, February 21, 1887

Was talking with Gilbert this morning about going to the Black Hills. He seems to think that I could go as well as not. He thought I could get a school there. I made up my mind if I could get or make sure of a school, I would go. It would open up a new world to me. I wrote a letter to F. L. Cook of Spearfish to see if he would give me any encouragement.

Friday, March 4, 1887

There were sixteen or seventeen at school. Fred Conner was there in the afternoon. I was sorry that he left school for I thought a good deal of him. I felt my heart go out to him and felt like clasping him by the hand but hardly spoke to him. How little others know what is in our own hearts.

Monday, April 18, 1887   At Mr. Battin's

Have been here most of the day hunting gophers; killed about 25. What I want now about the most of any book is a Botany as I wish to take up that study systematically this spring.

Friday, June 27, 1887

The last day of school. Have just returned from my first lecture, if I may call it so. Have often wished I could express my thoughts in public but have been too diffident to make the attempt before. I did better than I thought I could and it made me more confident in myself. The subject was the Effect of Alcohol on the Human System. There were but ten there.

Friday, October 28, 1887

Today ends the first 25 years of my life and it is a sad thought to me. I have lived a quarter of a century and as I look back on the years I have lived it seems like ages. As I think of earlier remembrances it seems these things took place way back in another world. I can remember, and it must have been some of the first gleams of memory that shine out from that eternity of nothingness before. When I yet slept with my parents I remember the white walls. I remember the school house and a hundred incidents connected with it; the long drives to weekly meetings, the long sermons and long meetings, my first reading, then my love of history and poetry and at last science. My aspiration to be a poet, an author, a writer, a scientist, a naturalist and through all a first-class school teacher.

Thursday, January 12, 1888   The snow storm

Wednesday night the wind blew a gale from the S. E. all night drifting the snow and sending it into the house. I went down to the schoolhouse in the morning and built a fire but no one came. I read and studied but along toward noon I began to think about going down to Starr's and see the folks, especially one whom I seemed to like better than the rest. . .Thought I would pretend to see about the next meeting etc. Oh, how deceitful we all are. I sincerely wished afterward that I had gone out, little realizing what was to happen. At noon the wind changed from the S. E. and came blowing furiously from the N. W. seeming to take all the snow with it. Soon I could not see more than two or three rods. The schoolhouse shook and the wind seemed bent on taking all before it. I read and studied and night came and no letup. I thought I would run out to the road and see if there was any prospect of getting home. I turned to go back and could see nothing but snow. I could hardly look up as the wet snow would plaster my face in a moment. I thought I was going right but if it had not been for some hitching posts I fear I would never have got to the schoolhouse.

Friday, January 13, 1888

Electricity - I was obliged to stay at the school house and carried enough coal to last all night. I read a good share of The Young Voyagers. Witnessed a peculiar phenomenon.

On touching or putting the knife near the stove sparks were emitted making a scraping noise. When held at a greater distance some times violet colored streams issued from the end of the knife. On moving it closer, bright flame took their place. When I took hold of the stove door I received a shock and sparks were seen.

The blizzard of 1888 is considered the worst in the recorded history of the Dakotas. Many people lost their lives in circumstances similar to those mentioned by my father. I have heard him tell of the incident many times. If it had not been for a couple of tethering posts he undoubtedly would have wandered, bewildered, on the prairie and perished, as he was only scantily clad for such severe weather. By bumping into the posts, while going in the opposite direction from the school house, he was able to orient himself and return to it. A letter from his mother in Medford comments on the storm as follows:

Dear Earl,

You don't know how thankful I am that you found the school house and did not have to perish in that awful storm, as many did. You must have had terrible feelings about that time how anxious we were to hear from you. It seemed a long while before we heard any thing. Everyone here that had friends in Dakota were waiting anxiously to hear from them. Well, I hope you can do better than to stay in Dakota another winter.

Is your asthma better this season? (you wrote to Dr. O'Leary). I don't believe that Dakota is the right place for you anyway.

from your Mother

At this time my father was delving ever deeper into science and he was also writing a great deal of poetry. It was a surprise to me, when transcribing the diaries, to find a couple of poems which I had always assumed were written many years later. One, The Great Unknown, which was read at his funeral in 1931, is recorded later.

The other, Nature's Noblemen, is here recorded under the date written. As the diary continues he was still feeling the effects of the storm.

> Saturday, January 14, 1888
>
> Have not delivered any of my lectures on physiology yet this winter. There has seemed to be something to hinder all the time. It is somewhat perplexing to know how to begin. What to take up first. I suppose there is a physical basis of life at the foundation, or near the foundation, to speak more correctly for I suppose the formation of protoplasm is only an event in the long line of succession which began we know not where and ends in that mysterious, complicated thinking arrangement, Man. We go downward from man to the cell and protoplasm - from protoplasm to the chemical elements and are lost. From chemical they came, how or why we know not. Of course we cannot get to the foundation but we should present these things so they may be best understood.
>
> Monday, January 16, 1888
>
> It seems that we almost live outside of the world in a little out of the way place of our own in a world of snows in these cold rough days. We hear little of the great world without, of its thoughts and feelings yet it sometimes visits us, the sufferings that follow man into every nook and corner of the broad earth, sometimes come to plague him here also. Yes oftener than not he is almost continually harassed. We hear occasional accounts of the terrible storm. One man (Mr. Gemer) who has been lost, they say, has been found dead. He started for a barrel of water it seems with his ox. The storm struck him. They found the ox alive I believe. It is reported that 11 have been found dead in Beadle County. If the storm went through to Texas as has been reported how many widows are mourning by cheerless hearths? If it extended south as far as reported it must have destroyed thousands of head of stock. The storm was not as long as some but I think for intensity it will be among the foremost as recorded.

Saturday, January 28, 1888

    Go forth into nature and see what she has to show thee. Enter the silent wood and lose thyself in thoughts unthought before. Let fancy construct worlds unknown - fairy worlds of the mind. All this is wonderful, but the wonder is of thyself the mystery of the mind and that matter can arrange itself, know to perceive, to perceive other forms, other arrangements of matter and then to think beyond, to construct a new world of its own yet of fragments of the old.

Wednesday, February 8, 1888

<u>Nature's Noblemen</u>
By Earl Douglass

They went out into Nature;
They left the traveled way
To read her deepest secrets
In the open light of day.

They fled to wildernesses,
Away from the beaten road,
To search the pathless mazes
Afar from man's abode.

They went out into Nature
With firm and joyous tread
To read in Truth's great volume
Whatever there was said.

They found the leaves were scattered,
With here and there a page;
And some were being written
And some were dim with age.

And there were wondrous stories
That never had been told,
Printed in rocky tablets–
Tales of the days of old.

There were some who, starting, faltered,
Or went with cautious tread
When they heard the low-browed scoffers
Who jeered at what they said.

And some shrank back with terror
From pathless wastes unknown,
And vast, untraveled forests
And unsailed oceans, lone.

But these the unknown tempted;
The wastes they did not fear;
Where was even in the deserts
An unseen presence near.

And they went, fearless, forward,
For when they but looked back
They saw in the ways of error
Only a blood-stained track.

They trusted Truth was safest –
That it could never die
But would conquer ghastly error
And triumph by and by.

They went through wildernesses
Where no man's feet had trod
To find out Nature's secrets –
The unfound way to God.

And so they searched and labored
To find a way from night
And sin and pain and sorrow
To Truth's and Freedom's height.

⌘ ⌘ ⌘

A START IN COLLEGE:
The University of South Dakota

Sometime in March 1888, my father made a choice between three possible plans for the future. He had thought seriously of going to the Black Hills of South Dakota, where he could pursue the studies of geology and mineralogy and perhaps even collect the bones of prehistoric animals; he could move near one of the lakes in the vicinity where he lived and spend a year, living as a hermit, to study the plants and animals and write; or he could enroll in college. After finding that travel to the Black Hills was difficult and there was no definite employment waiting for him there, he chose a term in college. As soon as his school was out he started for Vermillion and the University of South Dakota.

> Thursday, April 5, 1888
> Am now at the depot waiting for the train to take me to Vermillion.
> Went to meeting last evening and bade the friends good-bye. They have seemed to be friends indeed. There was one especially I felt peculiarly tender at parting with but probably will soon get over that. I think a good deal of her but do not think her the proper companion for me or I could be tempted to ask her to share her fate with me. Do not know if I shall return to Iroquois but probably will if I live.

> Friday, April 6, 1888
> Had a long wait at the train, as usual. . . . Came up to the University and saw the President. He seemed very kind and glad to welcome me as have all the professors I have met.

Have arrangements more satisfactory than I expected. Have a room provided with stove, table, bedstead, springs, mattress, and wash stand free of charge. Get board in the Ladies Dormitory for $2.00 per week.

Wednesday, April 11, 1888
Have about decided on my studies. They are Geology, English Grammar, English Literature, Botany and I think I will spend about a half an hour a day drawing. This will give me about all I can possibly do but I desire to gain all I can and learn all I can about teaching as well as obtain knowledge.

Sunday, April 15, 1888
Started out in the morning on my intended trip. Walked, I suppose as much as 17 or 18 miles. It made me rather weary. Think I was paid for my trouble perhaps. When I got to the mound (Spirit Mound) I lay down on the sunny side and wrote some verses. What a grand place it would be for some great teacher to address a large body of men, and a person can imagine the Indians thinking that the Great Spirit came down and the legend of the Pipestone Quarry was brought to my mind.

Monday, April 30, 1888
Attended a theatrical performance for the first time in my life. Had a prejudice against them, I must confess, but do not believe in taking other people's word altogether and thought I would sometime see for myself. Felt uneasy this evening and could not get my mind on my studies so I thought I would go down town and I might go in. Thought I would see the evil and would not go again. Was on my way home once and thought I would not go but finally turned around and went back. Did not like to spend the money and hardly knew whether it was right. I was very much pleased with the acting I must confess and it seemed to me about perfect.

Monday, May 7, 1888 – Rainy nearly all day.
Feel very happy for a mortal. Don't have much to trouble me. There is a solid comfort a person does not find anywhere else in living by oneself, as I do, with nothing to do but study. With plenty of oil and coal to burn and not feeling that it costs me anything. Oh, selfish humanity. Well, I would be willing to pay for such things if I had the money. Have been spending my time more on my lessons.

Thursday, June 7, 1888
Finished recitations. Do not expect to be examined in but one study, Botany.

June 11, 1888
Have packed part of my things. Have a box of books that I think I will leave here. Do not know whether I can come in the fall or not. It looks dark now about attending the winter term as I do not think I can earn money enough on the farm to carry me through two terms. Probably will have to teach.

A letter from his mother in Medford, Minnesota, written April 22, 1888 comments on his start in college:

Dear Earl,
I will try this afternoon to write you a few lines and let you know that we are still alive. Now Earl I am so glad and thankful that you have such a good place and such a good chance to go to school and are so nicely situated and that they are so friendly to you. . .I hope they will continue to be so and that you will continue to be worthy of true friends wherever you go. Everyone we have told here think it a good chance for you. . .Jenny says Earl will turn out to be a professor won't he?. . .try to take care of yourself and be a good boy.
                        from your mother

A letter from his cousin, Alice Carpenter, with whom a regular correspondence was kept for many years, reveals her feelings concerning his college venture.

> Sunday, May 2nd 1888
>
> Dear cousin,
>
> Your letter is gladly received and will take advantage of a few spare moments in answering it. . . . We heard through Nettie that you were attending the University at Vermillion quite a while before you wrote it. Suppose you will never be satisfied with yourself until you have reached the top of the ladder of knowledge, attained the highest of the attainable. I can tell you one thing though, Earl, you mustn't sacrifice your health for all the world of education, as anything without health is like a ship without a pilot. You once told me that, "advice rejected returned to the giver," now I can return the compliment. Your folks told mother that you were not very well and that you still have asthma and will have to be pretty careful of yourself.
>
> Your cousin
Alice

⌘ ⌘ ⌘

PEDDLING A BOOK:
A Sore Trial

The hope that he might continue his college education was now Father's predominant ambition. The small amount of money he could earn by working on the farms offered little hope for successfully financing his college attendance the ensuing fall. Since there were no schools to teach during the summer, his thoughts turned to other possibilities of making quick money. In short, he became a book salesman. He had received a letter from a book company which read, in part, as follows:

> Law, King & Law Publishing House
> Chicago, Illinois
> March 17 1888
>
> We have learned from reliable sources that you are a person of enterprise and energy...
>
> Many persons are entering our employ at the present time and we are prepared to give you an opportunity to learn our business and rise in it. In your case we will guarantee you a larger salary for the same amount of time than you can make by teaching.
>
> Law, King & Law

It is not clear how Father got in touch with this firm but apparently he answered a letter sent to him and was referred to a Mr. Musser, who had the Vermillion District. His offer of employment was accepted. Now it seemed his pursuit of knowledge depended on his success as a salesman of a book on home medicine.

Upon leaving Vermillion he went to Sioux Falls, South Dakota and attended Camp Meeting where he met a number of old friends from Minnesota. Among them was Will White, a brother in the church and missionary with whom he had been quite close, and who later assisted him in his struggle through college. After Camp Meeting, which lasted about three days, he journeyed by horse and buggy in the company of a friend to Iroquois.

> Tuesday, July 3, 1888
> We got to C. Gibson's at about sundown I guess it was. After supper I came up to Alfred Battin's. Expected to get some mail in Iroquois but the postmaster had forwarded it to Vermillion...I rather expected to get a letter from Mr. Musser. I am anxious to know what I am going to do.

> Wednesday, July 4, 1888
> I received a letter from Mr. Musser. He wants me to meet him at Wolsey. Wants me to engage in the business permanently. He says some of his men are making $250 per month. It is a little tempting to me if I can go to school. An education is worth more to me than $1000 per month. I do not think it best for me to engage permanently in such business.

> Saturday, July 7, 1888
> Went to S.S. and meeting...I have mentioned before a feeling toward a certain girl here. I find I have not lost the love I had for her. It seems sometimes that she will overcome my better judgment. As I see her now it is hard to resist the temptation to win her...Someway, something has blinded me to the woes of matrimony lately and it almost seems I would wed if I could get the right one...

> Thursday, July 12, 1888
> At Wolsey. ... Started on the 11:15 train. Stopped in Huron. ... There is some kind of a convention there to take the necessaries to become a State. It is rather aggravating to see how they act about admitting Dakota when she has been

fitted for statehood so long and fulfilled all the requirements. This is a small town of about 300 residents.

July 13, 1888

Mr. Musser came just before night. He told me how much their agents were making but didn't tell me what he wanted me to do until we went upstairs. I found it considerable different than I had anticipated. What he wanted me to do was to sell the book, "Hand Book of Practical Medicine". I was perplexed to know what to do.

This morning we talked again. He offered $2.75 per day and I would only need to make a $10.00 deposit. For the territory he would give me all of Beadle County. I did not have the money but if I were sure of the $2.75 per day I could borrow it.

Saturday, July 14, 1888

I talked with Arthur about buying one of his horses. He asked $112 for the one I wanted. I offered $100. We all talked the matter over and several of us thought a tricycle would be the cheapest.... I would like a bicycle but could not carry what I would want to. I have wanted a bicycle for years and maybe this will open the way to get one. The thing that worries me now is how to get some money.

Monday, July 16, 1888

Went down to Iroquois. Thought I would see what I could do about getting a tricycle to canvass with. I heard Dumbarton sold them. He told me that a bicycle was just what I wanted and could carry as much as 25 lbs. That is what I wanted and have been wanting and longing for and patiently hoping for years. They had a wheel there that is just my fit and I longed to take it but did not have the money.... I was fortunate enough to get $20 of Alton Gibson. ... A little before sundown I completed the bargain, paying $10 down and the remainder in six notes of $14 each to be paid one month apart. I did not have a very easy time coming from town as I was trying to learn to ride. Did not get any falls to hurt me but was tired.

Tuesday, July 17, 1888

Got up in the morning and practiced on my bicycle a while. Got so I could get onto it and get off too, but not very gracefully. Went to town with Arthur. After we returned, I took my wheel and went down to the N. E. road. Did not succeed in riding very much but got thrown several times. Stopped at Bro. Starr's and stayed all night. Enjoyed being near Sally very well. I am not much of a star gazer but do like to see her once in a while or oftener.

Wednesday, July 18, 1888

After breakfast went on the road and tried to ride. . . . got some tumbles. It seemed hard to work the pedals on a level road and I feared it was going to run harder than I supposed. I felt somewhat discouraged about it. I got so before night I could ride where it was decent going. Rode perhaps 30 rods at a time.

Thursday, July 19, 1888 – Hot

Went to Iroquois. Took my bicycle along and rode where the roads were good. Did not get my outfit. Have not more than one half the money to pay for it. I hope I shall get sometime so I will not have to borrow. If I, a single man, cannot support myself and have a little ahead it is pitiful. Well, I do not expect to save much and go to school. Money is very scarce. Nobody seems to have much. It seems to me if I had every debt paid I might be happy, that is if I had an education, but there might be something else to make me unhappy. In fact am quite happy anyway and do not have much to complain of.

Monday, July, 23, 1888

Am very tired this evening, as I have been riding on a bicycle. I find it hard to go up a hill against a wind. . . Went down south of Iroquois a ways and fell off my bicycle and off a bridge. It did not hurt me but bent one of my pedal shafts. Well, I think I have found where I can get money enough to get my outfit, which is at the depot I suppose. Gilbert said if I could find where I could get it he would back me, and Mr. Zimmerman said he would let me have it on those conditions. There is not much fun going around trying to borrow money and I thought as I went to the bank and asked Mr. Zimmerman for some, "I will be above this sometime."

Sunday, July 29, 1888
   Have got so I can ride my bicycle much better. Can keep it in one track most of the time. Got dumped a couple of times on account of the wind. Got my second header in a mud puddle. I am living an irregular life at present. Hope to settle down to business soon.

Monday, July 30, 1888 – tremendously hot.
   Went to Iroquois. Received a letter from E.C. Musser. . . .have been reading over the rules for canvassing etc. and think of the debts I have got to meet and almost despair. Heartily wish I had never heard of the business. It was only my overwhelming desire to get an education that ever induced me to do it.

Saturday, August 4, 1888 – a terrible hail storm.
   After the meeting it began to rain and after a while it hailed. The hail stones were not very large at first but after a while large ones began to fall. The largest I ever saw. We could see them flash like meteors through the sky falling in straight lines from the clouds. They fell with great force and I think would have knocked a man down if one had struck him on the head. Alfred measured one and it was between 6 and 7 inches around. They made noises like real stones when they struck the houses. The poor cattle ran in consternation and it seemed they would perish.

Sunday, August 5, 1888

Thought it would be best, as bad as the roads are, to get a horse and buggy if I can, to begin canvassing. I saw Ezra this eve. and he said I could have one of his. Think I will start out Tuesday. Am afraid the hail (storm yesterday) will hurt my business.

Tuesday, August 7, 1888 – 7 miles south of Huron.

Started canvassing trip this morning. Had to go to Mr. Warner's to get the horse and got the buggy of Mr. Burgess.

Wednesday, August 8, 1888

Have been at work hard all day. Have visited 10 or 11 families. The excuse is they have not money.

Thursday, August 9, 1888

I started from Mr. Rutan's this morning and went North and after some time came to a house and stopped to canvass. I thought the little girl looked like Mr. Rutan's and found I had come back to the same place. Mr. R. did not charge anything for keeping me.

Canvassed the following day. Winter, Stone Baldwin. . . . Tried to sell a book and get a place to stay all night but did not succeed. Everybody nearly seems to be poor. Hardly anybody has money if they tell the truth.

Thursday, August 16, 1888

Am now stopping on the prairie to let my horse eat. Have canvassed six persons. Charlie took the book and paid me the money or I do not know what I would have done. I do not see how I can pay for my bicycle and there are other debts that should be paid. There is no room for despair. I must get up and do my very best. My feelings the last few days have made me long, with such intense longing, to go to school. . . . It makes me sad to think life is passing away and I am not realizing its grand aspirations and hopes. But I will probably succeed yet and look back on these days as the more profitable of my life. Oh, may I be strong in these trying times and come out better and wiser than before.

Took only two orders today. Have been at it a good while but have had a long way between houses. I do not see how a man can take 17 orders in one day as Mr. Musser says one agent did. Have a good many things to learn yet. Am getting more enthusiastic over the book and think I can represent it better. One thing I do not find out is how to get money out of a person that has not got it.

Thursday, August 23, 1888

Stayed at Mr. Black's overnight. I was troubled considerably as I have had an itching sensation nights, which I feared might be something that would compel me to go home. ... I read the book what was said on itching etc. Had feared graybacks before and feared I had gotten them somewhere, so drove down to the slough out of sight, fed my horse, ate what little crackers I had and took off my clothes and gave them a good search. Failed to find anything. Washed myself and shirts as best I could. Forgot to bring along a change of underwear and dared not spend money for a new one. This is the life of a canvasser, but one can expect trouble anywhere.

Saturday, September 1, 1888

According to agreement went down to Iroquois with my bicycle. I had made up my mind if they or Mr. Dumbarton would take it back it would take a big load off my mind. I offered to let him take it back. He offered to sell it and give me whatever he could get over $55. He seemed very fair and I left thinking more of him than before.

Thursday, September 13, 1888

At Mr. Bycroft's south of Huron. Am learning all the time and find good people wherever I go . . .It is hard times and people are poor indeed. Many intend to leave Dakota and very many have left... In some places it has hailed year after year and I hardly know how people have lived.

Sunday, October 28, 1888
Am 26 years old today. More than a quarter of a century is gone and nothing accomplished yet of any value. Have hoped to make my life amount to something but have been disappointed so far. I realize if I am ever to amount to anything I will have to start soon. I have decided what I would like to do and what I would like to make of myself - a teacher and scientist.

Monday, November 5, 1888
Weather quite comfortable. Came up to Bro. Hay's. There are thousands of geese around the lake now... They made a terrible chattering. They are nearly all white geese. They pass by in thousands every spring and fall, generally stopping for a few days on their way.

Tuesday, November 6, 1888
Presidential Election. ... Have been in the depot hearing reports from the election. Harrison seems to be ahead at present. Hope he will get there but am afraid.

The summer canvassing venture that had begun with high hopes of earning enough money to start college in the fall, ended in debt. Quite a number of orders were taken but they were hardly enough to pay expenses. Although he had to postpone his college education, he did learn the trials of a salesman. Mr. Musser wanted him to continue, stating that he did well to sell any books in such a poverty-stricken area, and offered him a district that was more prosperous. He declined, however, and took a small school near Huron, which he started teaching November 10th for $35.00 per month.

⌘ ⌘ ⌘

BROOKINGS:
South Dakota Agricultural College

My father taught at the little school near Huron, South Dakota until February 28, 1889. A small shanty about a mile away from the school served as a bachelor quarters. It had not been recently occupied and was rent-free. Because he could economize in this way he soon began to pay the debts he had accumulated during the summer and by January he was thinking of college again. He made up his mind if he could finish the school term with $25.00 he would return to college. He decided to go to the South Dakota Agricultural College at Brookings, where he found he could work for 12 1/2¢ per hour, two hours a day, while attending school.

He arrived in Brookings March 6th. He had considerable trouble finding a place to stay but was finally accepted at one of the dormitories. He registered March 7th taking German, algebra, physics and botany. At this time botany became one of his great interests. A few entries in his diary recount college days in Brookings.

> Friday, March 8, 1889
> We have been fixing up our room today (room 18). Two of the boys have been papering and I have been painting this afternoon. Paid my deposit this afternoon ($10.00). Prof. Kerr says there will be plenty of work to do so I am in hopes I can stay all summer.
>
> I think I have very good roommates…Do not know how I will like the school but think I will learn something. Must learn to appreciate the privileges while I have them. Have longed and longed and almost languished to be in school and getting an education.

Wednesday, April 10, 1889

Have been examining trees this afternoon. Put in about three hours or less taking notes regarding their condition, etc. The Prof. was not well satisfied with it. He came down this afternoon and found they are nearer as I had reported than he thought

Friday, April 12, 1889

Feel rather tired this evening. Worked three hours this afternoon shoveling dirt. Prof. Kerr concluded he would not have me continue taking notes. Thought he would have Inffer do that as he knows more about trees. Well, I guess I can hoe and dig good and that is a consolation.

Thursday, May 9, 1889

Told Prof. Lilly that I wished to drop algebra. He said I would have to see the president. I saw him and he said he would see Prof. Lilly. He said others had been having trouble. I have not time to get the lessons I am sure.

Wednesday, May 15, 1889

Am not as well as usual. Got some tar and sugar and Frank took it down to the kitchen to make it into syrup. It was boiling and stunk Prof. H. out. He went down to the kitchen, a couple of the boys were there and Fred was stirring it. The Prof. put a stop to it. Fred brought some of it up. I took a little, so did Ira. He was sick all night and lay and groaned. I have had a mean feeling in my intestines ever since.

Saturday, May 18, 1889

At dormitory. Am coming down with the mumps it seems. My chops are swelling and somewhat sore.

Received a letter from Bro. White which lifted a great load from my mind. He is short for money but said he would send me $10.00 in a week or two and the rest as soon as he could get it. I do not know what I would do unless he helped me. I hope I may be grateful as long as I live.

Wednesday, May 29, 1889
Finished examination in physics, stood 85. Drew my pay for this month, $4.31. Paid $6.50 on board. Received a letter from Will White. He is trying to get money for me but has not succeeded. Think I will leave school. I terribly dislike to but see no way out of it now.

Monday, June 3, 1889
School began again this morning with a good attendance. Did not go to any classes as I do not know whether I shall attend this term or not.
Was talking to Prof. Duffey about getting a school. He said he hated to see me leave and offered to loan me $10.00. He knows how it is. He used to borrow money and had a hard time getting along.

Tuesday, June 4, 1889
Went down to the Court House to take teachers' examination. Began at 10:00 A.M. and did not get back until 6:00 P.M...Saw a pretty girl.

Friday, June 7, 1889
Went to my classes, geometry, chemistry and German.

Thursday, June 13, 1889
Worked 4 1/2 hours.
Received a letter from Ida; also my certificate from Mr. D. Robinson, the county superintendent. Average standing was 90.

Friday, July 5, 1889
Have received no word from Will White and if I do not tomorrow, I shall make different arrangements, will probably

leave here. Am in a discouraging, humiliating situation. Wish I could have known before, yet it may be for the best.

Monday, July 8, 1889

Went to see the president so as to get excused from the school. He wanted to know what was the matter and asked me some questions so I had to tell him the whole story. He wanted me to stay another week. He thought the money would come alright.

Wednesday, July 10, 1889

Received a card from Will White at last. He says it is just impossible for him to send me any money now. I am sorry I could have not known before. It seems our friendship has about ended. He is a noble man, and aside from his not writing as I think he ought, I feel the blame is all my own. I have caused him a great deal of trouble undoubtedly. I must depend on Mr. Douglass after this.

Thursday, July 11, 1889

Went to the president and told him what Will White wrote me. He wrote a note to Prof. Keefer asking him to listen to my story. He said he wanted to get up a college herbarium but had not the time himself. He said he would see the president. They decided to let me work four hours per day.

Thursday, July 18, 1889

I feel much better at present. Earn enough to pay my board, and enjoy the work. It is a pleasure instead of a drudgery.

Tuesday, July 30, 1889

Made another failure in geometry recitation and was angry with myself. My life is threatened to become a failure because I have learned the wrong methods of study. Have not been thorough and persevering enough. Well I am going to keep trying. I may win like the turtle if I keep on plodding.

Tuesday, August 20, 1889

Made my first record in athletics today and as there were only two contestants I skillfully won the second prize. Mr. Bacon and Mr. Patterson were to try for, or rather win, the prizes for the broad handspring but Mr. B. was sick so they wished me to enter. I did not want to but finally consented.

Friday, September 6, 1889

Some of the boys wanted me to go over to the election of officers for the Collegian and I was elected scientific editor. It will be a pleasure to do the duties of the office if I can do it credibly.

Saturday, September 7, 1889

In the evening prepared to go to Oakwood Lake. Mr. DeGoff wanted me to join the Philanthean Society. He said the girls wanted me to go, which was flattering, but I confess there was one I cared more about than any other and when I heard she was one of them I rejoiced greatly...

Monday, September 9, 1889

I am getting down to business. Getting things straightened up so I feel easier and can study better.

Began shorthand today. I intend to use it as fast as I learn it. Have begun an article for the Collegian on scientific books.

For all practical purposes the course in shorthand, mentioned on September 9th, interrupted Father's diary for some time. He soon started using it, a type of shorthand which is not commonly used today. Enough can be deciphered to tell that with the term that ended sometime in September, he left Brookings and again started teaching at a country school.

⌘ ⌘ ⌘

BOTANIST:
Mexico and St. Louis

A few written words and sentences, intermingled with his newly acquired shorthand, together with letters, reveal some of the events that followed after leaving Brookings. My father accepted a school near Cavour, a short distance from Huron, South Dakota, at $45.00 per month and, as usual, started settling the debts accumulated the previous year. He was deeply moved by the death of Professor Olson, the President of the University at Vermillion, where he had first attended college. The following was written, without the use of shorthand, soon after he received word of Olson's death.

> December 3, 1889
> It was with pain and sorrow that I heard last eve. of the death of Prof. Olson. He was my ideal of a man. I loved him as I had never loved another man perhaps and I had not associated him with death, or at least it seemed he had a great work to do and I thought he would live to do it. . . . It is not right that such a one should be taken away in the zenith of his strength. But oh, the good he might have done. The world has so few such lights it could ill spare him. This dark selfish world needs all the unselfishness it can get, all the love it can command.
> It seemed he had learned what few men ever learn -- how to live. His work was in the hearts and souls of men and such work endures. We miss him sadly, bitterly miss his bright hope, his unsullied life, his love, his example. Bitterly we regret circumstances were so cruel.

The entry of the last day in the year is written in longhand and touches on Father's religious problems.

> Tuesday, December 31, 1889
> It has been years since I have had a strong faith in the future life beyond the grave. ... I think the best we can do is make the world a little better or nobler for our being in it. ... One of the greatest things ever said about Christ was that he went about doing good. Oh, that we may follow in his example.
> I have broken away from the former ties to some extent. I so dread to give pain to my mother, father and sisters and friends but I felt I must if I would be honest. I wrote to the church to which I belonged to have my name taken from the church records.

Father's tentative plan for the year 1890 was to finish the school he was then teaching, go to Brookings in May for a term that would end in July, attend the National Teacher's Association in St. Paul, and finally to return to Cavour to resume teaching at the same school in the fall. This plan was followed only in part. The school term in Cavour was finished and he returned to Brookings to attend college until July 3rd. His greatest interest in college had now turned to botany. He had worked hard and with enthusiasm to prepare a college herbarium that received considerable recognition and was exhibited at the county fair.

It is not clear just when, but sometime during his stay at Brookings, he got in touch with a Mr. Pringle, a botanist, who was then in Mexico making a botanical collection. It was apparently through Mr. Duffey of Brookings that he was recommended to Mr. Pringle. The thought of venturing into a foreign land in pursuit of such an interesting study must have appealed to him for he apparently wrote to Mr. Pringle. On June 2nd he received a letter offering him a job as assistant at $25.00 per month and expenses. He accepted the position. A letter from Mr. Duffey, who had now moved to St. Louis and was employed at the Missouri Botanical Gardens, reads in part as follows:

July 14, 1890
My Dear Douglass,
   Your letter just received. I am glad to hear from you and glad to hear you are going to Mexico. I hope you will stand the climate and will come home well and tough. If the climate should ruin your health or kill you I could never forgive myself for sending you there. I did not think about your asthma when Mr. Pringle inquired. I hope the climate knocks that out of you and you come home as tough as a pine knot. ... I am sure you and Mr. P. will make a team and love each other.
Yours truly,
J. Duffey

   As soon as the term was completed at Brookings, Father went to Medford for a short visit with his folks and was soon on his way to Mexico. So much shorthand is used in the diary at this point that only a glimpse is given of the trip and stay in Mexico.
   The train was boarded in Medford July 15th. He passed through Kansas City on the 16th, was in Laredo, Texas the 17th and arrived in San Louis Potosi, Mexico on the 19th. From the little information available the trip and the experience of seeing new country was a delight. Pringle and he stayed in San Louis Potosi from July 17th to August 7th. During that time many plants, entirely new to him, were collected. The journey would have been much more pleasant had it not been for his poor health, which hastened his return home. He did, however, enjoy a short but interesting excursion to Mexico City.
   Father departed from San Louis Potosi August 7th and returned via Laredo to Little Rock, Arkansas arriving August 10th. From Little Rock he went to Memphis, Tennessee where he boarded a river boat for St Louis, Missouri. Mr. Duffey met him in St. Louis, August 14, 1890. He remained in St. Louis where he engaged in work at the Missouri Botanical Gardens under Professor Trealease until March 17, 1892 when he returned to Medford for a short stay.

Father's stay in St. Louis appears to have been rather pleasant. The botanical work was intensely interesting to him. At this time he almost turned completely to botany as the field for his life's work and study however the unexplained mysteries connected with paleontology and geology seemed to keep beckoning. Time was not completely lost in these fields while in St. Louis, however, for he had access to a vast store of knowledge. Considerable work and study was done in the museum at Washington University where he received some college credit. A few of the diary entries written during this period which were not written in shorthand follow.

>Tuesday, August 19, 1890
>Began mounting poison plants, quite a lot of them.
>Wednesday, August 27, 1890
>A letter from mother made me feel better, was anxious to hear from home. The folks were surprised to hear I was in U.S. Had read in Brookings paper I was in S. America. I almost wish it were so if I could keep my health.
>
>Thursday, September 24, 1891
>Another bad day. There is dust in the road, dust in the houses, dust on the grass and weeds, dust, dust everywhere. No rain, a dry hot sky over a dry hot earth. Even the clouds in the sky seem to be clouds of dust that have succeeded to get out of the dirt.
>The plants wither, the flower fadeth and man goeth to his long days work and the dirty faced boy goeth about the street.
>The good pray for rain and the bad swear because it is so dry.
>
>Wednesday, October 28, 1891
>This is my birthday. ... Feel as young as ever. ... Almost 30 years. Many hopes have gone and changed. ... learned some lessons perhaps, some hard lessons of life.

Tuesday, November 10, 1891
   Went to a banquet. . . . Banquet for Botanist.

Monday, December 7, 1891   At museum.
   Finished reading conquest of Mexico. . . . Would like some geological reports - Colorado, Utah, California.

Thursday, December 31, 1891
   This is the last evening in 1891, 365 days have passed and I lived through them all.

Monday, February 13, 1892
   Went to Gardens and got nearly everything straightened up. . . . Received letter written by Nettie, partly by mother. Nettie and Pa quite sick.

Wednesday, February 15, 1892
   Am now at Sparta, Wis. . . . train is waiting for breakfast. Started from Chicago at 11:55 last night.

⌘   ⌘   ⌘

COLLEGE UPRISING:
Expelled

After a short visit with his folks and old friends in Medford, Father proceeded to Brookings. There, if everything went as planned, it would be possible, after almost ten years of struggling, for him to continue his education and to graduate from college the following spring. He arrived in Brookings March 1, 1892 and was well-pleased with the job he was given of rearranging and improving the museum.

> Wednesday, March 16, 1892
> Began work in the Geological and Mineralogical Museum. This is where I want to be now and there is a good deal of work to do evidently. Everything is confusion and has got to be arranged and most named.

> Saturday, May 14, 1892
> Am still at work in the museum straightening up and getting things in order. It begins to look more museum like.

In June the Ringling Brothers Circus came to town:
> Ringling Bros. Show was here today and I presume Brookings never had so many people and horses and wagons in it at one time. From every direction they came with buggies and lumber wagons with sweethearts and families to see the great show...
> Feel discouraged and disheartened tonight.

During a break in college between June 19 and July 11 Father and his sister, Nettie, who had been staying with him in Brookings, together with some of their friends took a trip by horse and buggy into northwestern Minnesota to visit the Beard family. It was primarily a pleasure trip and an enjoyable one. They camped and did a good deal of fishing and he mentioned seeing Cora Beard, with whom he had his early love affair. College resumed August 20th. The following are a few entries from the diary touch on college activities.

>Friday, September 16, 1892   another fine day
>Had a meeting of the Collegian staff this afternoon and dismissed some important matters pertaining to the Collegian.

>Saturday, October 15, 1892
>Dug potatoes for a while and at about 10:00 a.m. got ready to go down to have pictures taken with the rest of the Collegian staff.

>Thursday, October 20, 1892
>We had a meeting of a committee of 20 that had been appointed for the purpose of seeing about publishing a volume of poems written by present and former students. . . I think we can get up a very respectable volume.

>Tuesday, November 8, 1892 - a very cold night last night
>The faculty decided not to encourage the publication of the proposed book of poetry. We are not ready to give up so easily now we have made up our minds to do it. I think the cost of the thing was very much underestimated but I believe we can get up a book that can pay for itself.

>Thursday, November 10, 1892
>The last day of school for this year. Attended a party at Mr. Whitten's this evening.

Before leaving Brookings, to teach a small school near Huron, we find the first foreboding of the stormy period that was brewing at the college, and although the trouble had been in-process for some time, any thought of Father being personally involved was very remote. His plan for the future was to return to Brookings in the spring, after the school was finished. The following diary entries were made during this period.

> Monday, November 14, 1892
> Have begun to read from the report of the expedition to find a R.R. route to the Pacific Ocean in 1863-65. Want to find out all I can about this region. Have two objects in view, to write a poem and to find out about geology, etc.
>
> Wednesday, November 16, 1892
> The board of regents have been in session at Aberdeen and have outdone their old record. They have removed Profs. Keer, Aldrich, Orcutt, Laphtem. We have been talking over the matter pretty much since it was done and they say the whole town is indignant. There seems uniformity in the belief that Dr. McLouth is at the bottom of it.
>
> Thursday, November 17, 1892
> Nettie has gone again and I feel left alone. . .feel sad.
>
> Monday, November 21, 1892 cold, not so windy as yesterday
> My first day teaching for over two years. . .
> Be awake for time is passing
> And there's endless work to do
> Thou shalt fail but never falter
> Every day begins anew.
>
> Have thy aim and prepare noble
> And the sins that needs must be
> Shall be passing clouds that vanish
> And the distant voids fill.

Search to know for knowledge liveth
Tho the mind that fadeth dies
It may leave a grain of knowledge
That into the heavens rise.

And the mind that searcheth findeth
More than it had ever thought
And thou go ever onward
Moving on from thought to thought.

Monday, December 12, 1892

Got a letter from Prof. Williams telling of the death of Mr. Duffy. It startled me; I was not prepared for it.

Sunday, January 1, 1893

In beginning the new year it might be well to say something of how I began, though it would be a long story to tell all. My physical condition is by no means perfect. My lungs have been bothering me considerable and I have not been exactly well since I came here. I am improving though, I think.

I am still teaching a country school, my old occupation. I am sorry to say I have been digging away at books for fifteen years or more and have read very few and with all this pretending to be studious have learned very little, and am thorough in nothing. So now am teaching a country school and am not even fitted for that.

Wanted to accomplish a good deal this winter and have done practically nothing. I have sometimes thought I had some genius but that fond delusion is often dispelled by the sunlight of truth.

Went rabbit hunting today and saw one rabbit.

Sunday, February 5, 1893

It can hardly be said I live this winter... I try to teach during the school hours but do not make much of a success of it. Would like to be in a big city like Chicago or St. Louis so I could get stirred up and more than exist...

Thursday, February 9, 1893
They are going to investigate the Brookings College troubles at Pierre. The Governor, Sheldon, it seems is a McLouth man and appointed Sherman and others of McLouth's friends, five new members in all, to the board of regents, but the senate will not confirm the appointments until the matter is investigated. If after all this McLouth and his gang are retained I do not know whether I will return or not. But I think there will be a change. I hope so at least.

Monday, February 20, 1893
College affairs do not seem to be any better after the investigation...I should think the Dr. would resign. If he does not I think there must be something serious the matter with him. If he stays I imagine he will have to be rather careful after this or there will be trouble with the students. But I do not think he will stay long.

Friday, March 10, 1893
The last day of school.

Monday, March 13, 1893
Am getting anxious to get away as I have so much to do if I get through. I think I will make it though, if nothing happens.

Friday, March 17, 1893
Went to class for the first time. Eng. Lit., Psychology, Chemistry.
In the evening went to banquet given by returning professors. It was acknowledged a grand success. I think it was the proper thing and the purpose was a good one. I think it the first manifestation of college spirit in Brookings since

I have been here. We met first in new… Hall, had some music and addresses of welcome, John Maguire followed by Prof. Foster. After hearty shaking of hands, we repaired to the Jordan House where we had a very fine supper followed by toasts.

After the banquet several of us boys talked in Mr. Lusk's room until about 4 o'clock. Most of the talk now is college trouble. Had quite an indignation meeting. We were of various opinions and did not decide what to do. It is a bad affair. I wish I were through. Had another meeting of part of the seniors this evening. Decided to send a petition to the regents and trustees.

Wednesday, March 22, 1893

Our committee met this morning and finished our request to the regents and trustees. We met at the 4th hour but failed to unanimously adopt it.

I am now classified in Veterinary Science, Analysis, Physiology, and English Lit.

The old horse doctor who it is said might put Corey out, and threatened that he would, is here and I feel that we really ought to make it uncomfortable for him.

Friday, March 24, 1893

Went to the horse doctor's class. A miserably poor specimen of a teacher. What an outrageous shame it is that such a thing should be put in place of Dr. Corey.

Tuesday, March 28, 1893

Went to the office today to see the President about Political Economy and he called me into the secretary's room and talked with him about trouble and other things. Several of the students went to him in the morning to try and get out of Veterinary as they did not like the teacher. He is not an educated man and is simply a horse doctor and does not understand the English language. It is sickening.

Thursday, March 30, 1893

Another critical time seems to have passed. We have been somewhat expecting that Mr. Lusk would be expelled and a quite large number would go if he did but he did not get any discharge. It is impossible to tell how it will terminate.

Monday, April 3, 1893

Mr. Lusk was called before a committee consisting of Dr. McLouth, Prof. Shephard and Prof. Philips. The editorial in the Collegian was read sentence by sentence and commented on. I do not know what they are going to do about it.

Tuesday, April 4, 1893

Had another large meeting this evening to see about the proposed newspaper article. The committee reported submitting the article. A new committee was appointed to revise it and I was one of the committee. We worked at it until about half past ten tonight but did not get more than a third of the work done. Mr. Liev, Mr. Mitgar, Mr. Wallace were members of the committee. I think there will be something done sometime if we keep on. We expect to be expelled, that is we feel we are likely to go, if we have any such thing printed but we are cowards if we do not heroically stand up for right.

Wednesday, April 5, 1893

Did not come home but stayed over night with Aylsworth's. In the afternoon we worked on the article we were correcting and got it ready to be read at a meeting in the evening. The report was accepted with a few little alleviations.

Thursday, April 6

Had a mass meeting this evening. There was some discussion, mostly with considerable warmth. It seemed that some who were against the action came with the design to break up the thing and make it fall flat. At last the President came also and wanted to speak.

Friday, April 7, 1893
Have got 90 signers to the article. Took a copy down and telegraphed it to the Sioux Falls Argus Leader and I suppose it will appear in the morning paper.

Saturday, April 8, 1893
They had faculty meetings nearly all day and a good share of the night. Messrs Menzer, Hoy, Aylsworth, Talcott, Philips; Philip and John Maguire, Wallace, Yoger, and myself were called up and all but Yoger, Wallace and Hoy were fired or suspended and were to leave by Monday night.

Sunday, April 9, 1893
Went to meeting at the Methodist church in forenoon.
Took dinner with Prof. Williams and wife. We look at things a little different but I believe he does the best he knows how. I know it is hard for him to sit in judgment on a friend and believe he hated to vote against me. Those who have been suspended have been invited over to the club for supper. I think it is the first time I ever ate there. The girls are very strong on our side.

Monday, April 10, 1893
Only about 40 or 45 students attended chapel and classes today. About 100 have pledged themselves not to go to chapel or classes until the suspended are taken back. I think that will be a very long time.
The Argus Leader came out with our article today. It also had a good, sound editorial on our side or the tenor of it sounded in that direction. We have until tomorrow to get out. It will be a bad thing to lose our standings. It will put us way back and I am getting pretty old to go back very much.

Thursday, April 11, 1893

Another day of the fight. We have got ourselves into the trouble, at least we have helped the faculty do it, and we have got to fight it out. The separation between faculty and students is growing wider and wider it seems. The students are growing more and more determined. Shannon has come but we hope little from him.

Saturday, April 15, 1893

These are stirring times but I have not had much time to write. The amount to be written staggers me. A large number of students went away today. I am sorry we have to part but I have done so much parting that my heart is getting rather tough, at least at times.

We "Senior Outs" had our photos taken this morning. It is sad that any such thing should happen in a college but I believe it is the only cure. I cannot say I am sorry for the action I have taken on it. Sometimes the future seems a little dark and we may suffer for it but I believe the good that will come out of it will be more than the harm to us. The Governor spoke in the chapel this evening. It was kind of a noisy episode I thought.

Sunday, April 16, 1893

Walked over to Mr. Al Evans', took dinner there and stayed until 5 or 6 p.m. Carrie is home. Walked from there to Mr. Sim Evans and stayed over night.

Monday, April 17, 1893

Left Mr. Evans at about 1 o'clock p.m. Was late for the train but got a ride part of the way on a hand car. Walked the rest of the way. When I got to town I found that the boys, several of them, had seen the governor and changed his mind considerable. He asked the faculty to reinstate the students. They had a meeting this afternoon but did not do anything of the kind. The governor is getting into stormy water I think. . .a good share of the students have gone.

Wednesday, April 19, 1893
    Am getting ready to go but do not know which way I am going yet. Miss Doring and Miss Farr are at Mrs. Foster's, spent the evening here and had a pleasant time. Mr. Liev and Mr. Aylsworth are in town yet. There must be something like 70 or 80 attending college now. Mr. Egerton says 63.

Friday, April 21, 1893
    Got packed up at last and started for home. Saw Mrs. Williams before leaving. She feels badly about the trouble and the results. It is an unpleasant position in which to place a faculty. Some deserve no better treatment and some are getting their just deserts. It has been a remarkable uprising; I think it worthy of careful study. It has been as orderly as a thing of this kind could be conducted.

There was a wide difference of opinion concerning the cause of the college trouble in Brookings. The different versions of the affair were strongly expressed in the various newspapers. One version was that it was a battle between two factions within the faculty, the one faction instigating the uprising and using the students as a means to get revenge on the other faction. The students strongly denied this view, claiming it was originated by them in protest of bringing politics into the school. Their main objection being the qualifications of the professors who were replacing those being removed.

The following is a quotation from the article published in the Argus Leader for which the seven students were suspended. Several of the replaced teachers are discussed in a similar manner in the article.

    "Mr. Chilcott, the new teacher of agriculture, is woefully ignorant of the subject he is trying to teach. . .This statement is made deliberately. The students in question are fully prepared to sustain it. . .Professor Foster, who was removed to give place to Chilcott, is a master of scientific agriculture and is now president of Montana Agricultural College. Mr. Chilcott was a member of the senate investigating committee, which recommended the removal of Prof. Foster. Straightway Chilcott was appointed to succeed the latter. Was this a coincidence?"

Thus ended my father's college days at Brookings, which were much different than what had been anticipated. Instead of graduation, a college degree, and a chance to obtain his first decent job, my father faced a dark future without a degree, broke and in debt with no place to go but home. The seven expelled students wasted little time lamenting their fate, however, but began anew to search for ways and means to continue their education.

⌘ ⌘ ⌘

11

AMES, IOWA:
A Degree at Last

As soon as the outcome of the Brookings upheaval was known, brothers, Philip and John Maguire, two of the seven students expelled, were on their way to Ames, Iowa and the State University. Mr. Lusk soon followed them. My father hoped that some way might be found so he could continue his college work making graduation still possible in 1893. He returned to his home in Medford and awaited word from Iowa. On April 26th he wrote, "Got a letter from Lusk which is very encouraging." By May 3rd, he too was in Ames getting classified. His subjects were: history, psychology, Development of the United States, American Literature, and geometry. Considerable trouble was experienced in getting their credentials from Brookings and a number of extra subjects were required, but eventually all was straightened out and all that remained between them and their coveted degrees was to pass the specified courses.

Six students went from Brookings to Iowa. Four from the originally expelled seven: Lusk, John and Philip Maguire, and father along with two girls, Miss Downing, Miss Farr who were among the ninety or more students who quit school in sympathy with the uprising. Father wrote in his diary of these troublesome and sometimes interesting times.

> Saturday, May 6, 1893 - a very fine day
> Mr. Lusk and I went on a geological expedition down to the R.R. bridge east of here and then went up the Skunk River. Found an outcrop of rocks on a creek near where it flows into the river. We found a good many fossils... 3 or 4 Brachiopods.

Tuesday, May 30, 1893
   Had a headache and fever during the night and thought I was coming down with something.
   The I.A.C. boys beat Drake University 15 to 2 in baseball.

Monday, June 4, 1893
   I do not very often stop to look over the work I have to do to get through but have been looking it over a little tonight and it almost staggered me. My case is very doughtful. I have got to pass 9 weeks of trigonometry and it is very doubtful about my doing that. Then I have to take examinations in algebra. . . . If I could get trig. though, I guess I would pull through.

Tuesday, June 6, 1893
   I am getting older all the time but do not know that I grow less foolish or less apt to fall in love or get over it less easily. Yesterday I had come to the conclusion that I did not care for any except Miss Duffey and then I had a little feeling akin to affection for Miss Ryan. . .

The spring term ended June 21st and the next term did not start until the latter part of July. Father was invited to the C. Wilson residence at Gilbert Station a few miles out in the country north of Ames. He had apparently become acquainted with the Wilsons through their daughter Jessie who was attending college. Miss Downing was also there. He had engaged a Mr. Brown at the college to help him through the dreaded trigonometry and was studying it along with playing a number of games of croquet with the girls while vacationing at Wilsons'. He writes about this in his diaries.

Friday, June 23, 1893
   Began trigonometry but did not get as much of it as I ought to. Played several games of croquet. Have got to do different from this or I am going to fail.

Friday, July 21, 1893

Left Mr. Wilson's and came down on the 6 o'clock train. Have had a very pleasant vacation. Mr. Wilson's people seem to be excellent people. Jessie told me to come up and bring my friends. She is just a lovely little being, though I did not get fully acquainted with her.

Tuesday, July 25, 1893 - hot as usual.

Think I will take Astronomy, History of Civilization and Economic Geology.

Mr. Lusk and I have been out on a rye stack talking about love and other important matters. He wants to go to another school and get his M.S. So do I but fear I will not be able to on account of money.

Monday, July 31, 1893

Received an invitation to write something in the line of poetry in the annual.

Saturday, August 12, 1893

Mr. Lusk and I went up Squaw Creek this afternoon. Had his rifle and fish line and hooks. Caught some minnows (chubs and shiners). Roasted them by a fire also some corn we got in a field. Lusk did not like the fare very well so we came home to supper. Mr. Lusk and I have been thinking some of going to the World's Fair the last few days.

Wednesday, August 16, 1893

Went to classes all day. Started on 10:27 p.m. train for Chicago Fair... Round trip $11.80.

Thursday, August 17, 1893

Got to Chicago about 8 a.m. Took breakfast and took a State Street cable and came down to the World's Fair. It is, "the biggest show on earth.".. such a wilderness of buildings and some of them enormous ones too. Went through part of the Arts Museum and Liberal Arts…visited the Anthropology building.

Saturday, September 9. 1893
Feel a little blue tonight for several reasons. Most of them I do not think very serious…out of money but may be able to get some after awhile.

Friday, September 15, 1893
Received $25.00 in a letter from Mrs. Duffey. She is going to St. Louis soon.

Saturday, September 16, 1893
Went up to finish examination in Trigonometry and answered only 3 questions. Pretty near went mad over one I thought I could do easily. Must have worked over an hour on it.
Went out surveying. Our party surveyed a line in the middle of the road south of the college.

Friday, September 22, 1893
Studied like a good fellow in the forenoon but went off this afternoon with the gun and did not get back until nearly 6 o'clock… Had a nice walk. Shot away about 50 rounds of ammunition. I've been moralizing some. Have been thinking what a man is. He gets some ammunition and a gun and goes around shooting at every living thing he is close to and for no purpose only for a little sport. Strange sport. I shot an innocent turtle that would hurt nobody and I do not think he had any particular desire to die.

Saturday, September 30, 1893
Last evening I went to college to finish my Trigonometry exams. It was dark, muddy and rainy and once or twice I almost concluded to give up. Told Roberts I had answered 8 of 10 questions but was afraid I had failed. I was exceed-

ingly glad to find I could get through without the last two but went to work on them. I have had a great time over Trig. and it has been the source of many fears and anxieties.

Friday, October 13, 1893
 Finished my thesis and took it to Prof. Byer. Do not know that I have anything rushing on hand. It is about the first time I have not had back or other urgent work since I have been here. It has compelled me to work at about full capacity. Just at the moment I have no very heavy burden on my shoulders.

Saturday, October 28, 1893 - my 31st Birthday.
 Mr. Lusk, Miss Smith, Miss Ella Wheller and I went to Gilbert Station on the 11:20 train.

Monday, November 6, 1893
 After months of suspense, doubts and fears my case is decided and at last I am to graduate, and what I have wished for and prayed for more than 10 years is on the point of fulfillment. . . . In the bottom of my being I am glad it is so and I would not for a thousand have it otherwise yet I have no great happiness, no marvelous exuberance of spirit. My experience as a student has been something unusual but I have had a never failing desire that has borne me through at last. There have been months and years of suspense and waiting and hoping and disappointment. There have been times when the heavens looked dark. But I can credit myself for only part of the victory. Had it not been for my friends I never could have done it.

Wednesday, November 8, 1893
 This was the last, the day we long had sought. It surely marks quite a point in our lives. It has been a long struggle and it makes me almost sick to recall the struggles I have been through to become a college graduate.

The graduation exercises began at 9 o'clock and they were long, dry, and tiresome but at last the time came when we stepped up and received our degrees and diplomas. There were sixty-one diplomas presented. We six from Dakota all graduated. Mr. C. Wilson asked me to go home with them and I finally concluded to go.

Monday, November 13, 1893
Went down to Ames to my boarding place on the freight, nailed up my boxes, finished packing and bade the folks goodbye... got my class picture, that I framed, and went to the depot. John and Philip were there to bid me goodbye.

Tuesday, November 14, 1893
Got home about half past four. Nettie was here. Ma and Pa gone to Tracey to see Uncle Chauncey's folks.

⌘ ⌘ ⌘

12

WESTWARD BOUND:
Montana Fossil Beds

The satisfaction of finally obtaining a college degree was practically obliterated during my father's stay in Medford after returning from Ames. This period of discouragement and anguish surpassed anything he had previously experienced. He was in debt and had completed college at a time when the opportunity for getting a decent teaching position was limited. Although he made application to a number of schools, the jobs were either already spoken for or the salary too low to meet his needs, or to satisfy him, for he felt that the salary should be commensurate with his qualifications. Of his financial plight he wrote the following in his diary.

> Wednesday, January 3, 1894
> I had, at the beginning of this year, 13¢ and debts of about $220 and am not working or earning a cent. This certainly is an "encouraging" condition of affairs. Am not well but am feeling better. So the year 1894 does not start bright for me.

> Sunday, January 14, 1894
> I am in exceedingly poor circumstances; the worst I have been for years and there is little hope of things being better for a good while.

At this point Father did not have money with which to purchase a book for his diary but he did not discontinue it on that account. Daily entries were recorded on partially filled pages of his 1893 diary. They were at times written in shorthand and are hard to interpret.

To keep his mind from his discouraging predicament he wrote, with the hope of having something published that might bring in a little money. He mentioned sending a story to a publisher who was ready to publish it, but it was necessary for him to put up $25 so it was never published. When not writing much of his time was spent visiting the few rock exposures that were present in the country around Medford and on occasion he found fossil shells. Frequent visits were made to Lindersmith's Stone Quarry. Mr. Lindersmith was cooperative and when something unusual was found, he sent word to Father.

He again wrote of his financial condition: "Oh, I am sick of living a life of being in debt and not a dollar of my own. It is horrible, when shall I be delivered?"

> Wednesday, January 17, 1894
> Soon after I arose Mr. Hatch, insurance agent from Owatonna, came to see me. He was here several hours and explained what I wished to know and it resulted in my making application for a policy of $2000, $500 to secure my creditors and the remainder to be divided equally between my father and mother. . . I gave my note for 60 days to be returned if I am not accepted. The only thing I fear in my individual examination is my heart. I have a little suspicion of that.

Father was later accepted by the insurance company. He borrowed $25 from a Mr. Lee in Medford on a six-months note. This relieved the situation temporarily. During the early part of April he received a letter from the county superintendent of Walch County, North Dakota, stating that it was almost certain that if he would go there he could get a school at $45 per month. By this time he was ready to accept almost anything. His parents and Nettie tried to persuade him not to go, hoping that he would eventually find something close to Medford. A little dissension, that had not been present before, was beginning to develop in the Douglass family. He was on the verge of leaving for North Dakota, regardless of the opinions of others, when he received a letter from Professor Foster, who had been discharged from the college at Brookings and had accepted a position at Montana State Agricultural College, just recently established at Bozeman, Montana. He had apparently written to Professor Foster and receipt of the letter changed everything.

> Friday, April 13, 1894
> Yesterday I received a letter from Prof. Foster saying that he had a school for me at $35 and board in the best geological region in the state. This morning I telegraphed him accepting the position and fixing the 24th as the day to start. The only thing that worries me is money. I have to get $20 or $25 more.
>
> Monday, April 23, 1894
> In the morning I took Dan and the buggy and took my trunk and satchel up to the depot in Medford. Ida, Alfred and family, including old Mr. Battin, came. I find parting with my family always unpleasant.
> Started from Medford on the 8:20 a.m. train. Had been anxious to get started and was glad to get away. My stay in Medford has been a sore trial to me and I have done nothing to earn a cent.

When he arrived in Minneapolis he stopped over a day to visit some college friends. He also took some of the fossils he had collected from Lindersmith's Stone Quarry to the University of Minnesota to try and ascertain their identity. On the morning of April 25th he was "heading west."

> Wednesday, April 25, 1894
> At Jamestown just as it was light this morning. It has been a great day for me. I never saw such scenery before, I think it much the best I have ever seen. Since leaving Bismarck it has been a great show. It almost seems as if today had almost paid me for the trip, that is if the cash had been mine. Near where we entered the Badlands was perhaps as interesting

as any A great wilderness of red conical hills. . .I would like to stop among them.

We are now on the Yellowstone River sidetracked at Terry and did not start until after eleven, then came to Miles City where they say they have the Coxey men corralled, 700 in number on 30 passenger coaches. At Forsythe the Supt. of the division is having the tracks torn up and emptying water tanks. It is said that three men were killed and yesterday's train has not got through yet so I am glad I stopped in Minneapolis. (A contingent of "Coxey's Army" of unemployed was led into Washington D.C. on April 29, 1894)

It is now about one o'clock in the morning and we are pegging along at a good speed. They say the track is clear now. It is reported that they are going to take the prisoners to Helena.

Thursday, April 26, 1894

When I awoke in the morning we were going along the high bluffs of the Yellowstone and my heart was joyful. I wish I had a picture of the scenery. It was grand. Cliffs of grayish rock arose abruptly to considerable height. As we approached the Big Horn I thought of the poem of Longfellow's relating to Custer's last battle. . .and I thought that near the junction with Yellowstone must be the scene of Custer's last fight. I found afterwards that it was not far from there but to the south.

I got quite a little acquainted with my fellow travelers and liked them better the longer we were together. Part of them, I judge, were cowboys.

Arrived in Bozeman just before noon. . .After dinner started for P.O. and to my great surprise met Mr. Wisner, an old Brookings College boy. He was in uniform and with others had been called out as he might be needed in the Coxey trouble. He immediately conducted me to the residence of Prof. Foster.

Prof. Foster and I went down town and I had the pleasure of meeting the county superintendent, Mr. Kay, the principal of the grammar department of the high school, the president of the college, one or two of the professors and others.

The future brightened. Past humiliation and depression faded as prospects for a job in fossil country increased. Father had come to a turn in the road. His vision had led him into the hills, among the sandstone cliffs, where his dreams were to be eventually fulfilled.

⌘ ⌘ ⌘

13

AN IDEAL COUNTRY SCHOOL:
Fossils at Last

The little country school about which Professor Foster had written was in the best fossil collecting area in Montana, a few miles up the Madison River from the Three Forks of the Missouri. This is the area where Lewis and Clark camped the summer of 1805 and named the rivers the Jefferson, the Madison and the Gallatin. The Madison is the middle of the three rivers and runs in a northerly direction through a valley for a number of miles before uniting with the Jefferson and the Gallatin to form the Missouri. The school was located in the Madison Valley where it is a couple of miles wide with bluffs a few hundred feet high rising quite abruptly on both sides. It was not commonly reached from the town of Three Forks, as might be assumed, but from the town of Logan on the Gallatin River to the east. From Logan the road cut across the hills for a few miles then dropped into the valley near the school.

Before proceeding to his school it was necessary for my father to take a teacher's examination and obtain a certificate. This was done in Bozeman on the 28th and 29th of April. His school was to start the 30th. He was anxious to get to his destination and start collecting fossils, which he imagined were numerous and just waiting to be picked up. The diary relates his first real fossil collecting experiences.

> Sunday, April 29, 1894
> Got up early in the morning, ate lunch and Prof. Foster and I went down to the street car and went down to the depot to take the 5:38 a.m. train but it was late and I had to wait quite a long time.

When I got to Logan I left my things and went over to the mountains north of the Gallatin. Found a few layers of rock with fossils but most of them were not in good condition. Walked a long way up and down over rocks and was a little disappointed in my find but got some things. It is a good place to study geology if not paleontology.

I have sent for a pamphlet by Peale on Geology of the region. Found Mr. Hutchison and rode home with him.

Monday, April 30, 1894

The first day of school. . .had seven pupils. I think I shall like the school very much. After supper I went over to one of the lower bluffs and found some fossils in a kind of white lime rock. I thought it was cretaceous but do not know. There are bluffs beyond that I want to get to soon.

Tuesday, May 1, 1894

Snowed in the night and nearly all forenoon and it did not go off until afternoon. After school I went east to the bluffs. Was somewhat disappointed in not finding fossils. Found one piece that seemed to be bone.

Wednesday, May 2, 1894

A cold day, at least the wind was uncomfortable. There has been no nice weather since I have been here. The winds cannot get here from any direction without blowing over the snow on the mountains so they are cold no matter which direction they come from.

Thursday, May 3, 1894 - Windy and disagreeable

Getting along very well. Have a cold and have had one ever since I have been here. Went out on the hills just north of here and found outcrops of rocks but no fossils. This country does not seem to be very rich in fossils but I think I will find some after a while.

The search for fossils continued almost daily but little was found except fragments and petrified wood until May 12th.

Saturday, May 12, 1894
After breakfast I took a lunch and went to the bluffs east of here. Thought I might find something along the "Big Gulch". At last I found bone in the rock instead of only finding small pieces lying loose. I was going along the side of the bluff where it was nearly covered with rubbish from rock above when, looking at a projection of conglomerate, I saw something that I thought looked like bone. I went to it and found I had not been mistaken. It was quite a discovery to me. At least I thought I had found where to find bones. When I had traced fragments of bone up the bluff they seemed to often end in a layer of conglomerate.

I went to work and carefully removed the bone by chipping off the rock and digging around it. It took a long time, and not knowing the shape of the bone, I broke off a part of the joint end. I think it is the arm bone, the bone just below the shoulder blade. I see now that I need to study zoology considerable and especially osteology. It seems almost certain when I think of it, that the bones I have found are too much like modern ones for cretaceous.

Sunday, May 13, 1894
Mr. Hutchison hitched a team to the wagon and the whole family and I went down to the river to see if we could catch some fish. It is quite pleasant along the river. There are willows, gooseberries, flowering currents, cottonwoods and I believe cherries. The river is very swift and quite broad. It was too riley to catch fish. I longed for a saddle pony so I could go on up the river. I need one very much.

Wednesday, May 16, 1894
After school I went hunting specimens northeast of Mr. Dell's. Mr. Dell's children, Francis and Delano, went with

me. Found part of a lower jaw with 6 double teeth. Stayed at Mr. Dell's all night.

Boarding the teacher, which was part of the contract of teaching in the Lower Madison Valley school in those days, was shared by the parents of the children attending the school. There were four families so the teacher moved from ranch to ranch, usually spending a week at each place. This made a rather nomadic life for the school master but it seemed to work quite satisfactorily.

> Tuesday, May 22, 1894
> Received from Washington, Bulletin No. 29 of the U.S. Geological Survey and the Fifth Annual Report. In the Fifth Annual Report is a map representing the geological areas of the U.S., which is a thing I much need. I find I have struck a richer place here than I supposed I had. Of the formations represented on the map all but the Eocene and Silurian are represented here in a compass of 50 or 60 miles. It makes me want to stay here another year.

> Wednesday, May 23, 1894
> After school I went home with Delano and Francis. After supper we went to the hills. At the foot of the rocks we found a rattlesnake which crouched under a rock and my efforts to dislodge him were futile. He did a good deal of rattling and stuck his head out and ran his tongue out at me.

> Sunday, May 27, 1894
> Almost every day I hear of places where I want to go but cannot as I have no means of conveyance. I want to spend several months, or years perhaps, exploring the country around here.

> Saturday, June 2, 1894
> Delano and I hitched Duke to a road cart and started for Logan in the morning. I took my express order $34.85 to the

express office and bought one ($19.00) and sent to Mrs. Duffey and one ($18.00) and sent it to Lusk.

Between 10:00 and 11:00 a.m. Delano and I started on a geological trip across the river. Went to the source of the Missouri and went down the river a little way, then ascended the rocky bluff to its highest point and surveyed the surrounding country. To the south is the flat green valley of the Madison with its bluffs on either side. From the southwest came the Jefferson and from the southeast the Gallatin Valley. A great deal of the flat at the junction of the three rivers, and for quite a distance up the Madison Valley, was covered with water. The rivers are very high.

After hunting for quite a while I found some good fossils but Delano was getting anxious to go and I had to hurry. Got a good jaw.

Sunday, June 3, 1894

I dug quite a while where Delano and I got the jaw. Tried to follow the formation around and finally found a place where there were many pieces of bone and teeth. I thought I had struck something rich. I dug and kept finding bones. Finally I exposed quite a number of vertebrae. Found the head but there were few pieces that did not crumble. . so with the tusk which was broken into thousands of pieces. Found the head of a bone of a hind leg. I think it was 10 or 12 inches across.

Monday, June 11, 1894

Have organized a botany class of four and we expect to study botany two or three days a week after school. The members of the class are Francis Dell, Sarah Hyde, Rosalie Hoellein and Margaret Hutchinson.

Sunday, June 17, 1894

Got on the morning train and went down to Three Forks. I had to wade to get to the edge of the town and then could not get farther. Found a place where I could get my breakfast. The new building is in sight where the son and daughter of Mr. and Mrs. Ryan were drowned about the time I came here. They found the girl about a week ago down below the great falls. It was very sad.

Saturday, June 30, 1894
    Went to Mr. Dells then followed an old road nearly east of Miss Campbells to the bluffs. The hill N.E. of Miss Campbells forms quite a landmark and with a round top projects out toward the valley like a promontory. I will, to distinguish it, call it Big Round Top.

When transcribing this portion of the diary I casually pictured the hill my father named Big Round Top as an ordinary hill with a symmetrically round top when viewed from the Madison Valley to the west. Although I had been through this general area many times I had never bothered to take the short side trip that would take me through this part of the Lower Madison Valley.

While on my way from Virginia City to Bozeman in 1964 I decided I would turn to my left where the road to Bozeman leaves the Madison River and follow the road into Logan and on to Bozeman. This took me through the area my father was talking about above and I wondered if by chance I might be able to identify Big Round Top. After about a mile the road rounded a sharp point and suddenly the whole valley came into view. When I glanced over its general features I suddenly realized that I was looking at Big Round Top. Although it was several miles away, there was no doubt in my mind of its identity but instead of being round on top, as I had it pictured, it was quite flat. Its roundness is on the horizontal plane instead of the vertical plane as I had thought. As I continued I had little trouble locating several of the other places mentioned in the diary.

Sunday, July 1, 1894
    Sallie and Frieda Hoellein and I went fishing this forenoon and caught 19 fish, mostly "whitefish".

Have been studying Leidy's Tertiary Vertebrates some... am getting considerably interested in the vertebrates and Tertiary formations.

The trains on the N.P. are completely stopped on account of the strike and we are completely shut off from the rest of the world.

Friday, July 6, 1894

Mr. Hoellein is talking of going to Helena; perhaps I will go with him if he goes. Trains do not run and they want him to take a load of eggs.

Saturday, July 7, 1894

There has been bloodshed in Chicago, smashing cars, burning R.R. property, according to reports. It must be terrible if there are no trains running in Chicago. People must starve unless there is a change before long. There was bloodshed in Butte between the P.A.'s and the Catholics on the 4th of July, one or two were killed. These are important times. It is liable to grow worse instead of better.

Brought home the fragments and the humerus, radius and ulna of the mastodon.

Monday, July 9, 1894

The superintendent visited the school in the forenoon. He seemed quite well pleased with the work the children were doing and with the interest they showed.

After school I went down to Mr. Hyde's with Sarah, took Buttons, Lorrie's pony, and went up to Mr. Crowley's gold mine. I went down in the mine and all through it. They are digging at two levels and have got quite a distance in both directions. The ore, they claim, is very rich and I do not doubt it. It is in Archean rocks.

Wednesday, July 11, 1894

We started in the morning in quite a rain for Helena. It rained until after we got to Logan, sometimes quite hard and was heavy going until perhaps the middle of the afternoon. We took lunch in Three Forks. I saw a man there who had a knife blade which he and others believed belonged to the Lewis and Clark expedition.

Friday, July 13, 1894

Came into Helena in the morning and spent the day walking the streets making a few purchases. In the afternoon I had quite a long talk with the state superintendent of public instruction, Mr. Steers. He is a young man and I enjoyed talking with him. He is studying geology, taking post graduate courses in Wis. University. I learned considerable geology from him. I find that there are also lake beds in the Jefferson Valley and along rivers which flow into it. Mr. Steers discovered part of a skull and limb bones of a horse near Dillon. He says there were other bones he did not get. I am anxious now to get into that region.

Sunday, July 15, 1894

On our return we lost our bedding and Mr. Hoellein had to go back about 3 miles. We unhitched at the Jefferson River beyond Three Forks but the mosquitoes were so bad that we pulled up and left. Camped in the hills near Logan.

Since taking the trip to Helena I am more anxious than ever to be so I can travel around the country and especially to make a thorough study of these lake beds. I am more convinced they have not been "worked up". I want to get more geological literature and devote even more time to geological study. I believe there is a rich field here and I might as well reap the harvest as anyone. At any rate it will be a good study in geology.

Wednesday, July 18, 1894

After school I went over to the bluffs east of Mr. Harris's farm hunting specimens and was very successful. I found first the side of a lower jaw of some animal in fine sandrock; got it out entire. Found also a large turtle which measured two feet when exposed in the face of the cliff. Found another turtle on his back and part, at least, of the bones are inside the shell. Found many fragments of turtles besides also perhaps a half dozen teeth.

On the side of the sloping bluff found the socket joint and pelvis and other fragments of another animal but did not succeed in finding where the animal was buried. It seems a little strange when I think how little I found at first and how much I find now. I am growing more and more successful. I thought I had about searched the bluffs once but at first I seemed to have had the faculty of getting along the hills and missing most of the fossils. But I am learning more where to look. It seems nearly every time I go out that I have pretty well exhausted the supply but keep finding more.

Saturday, July 21, 1894

Took Puss, Mr. Hoellein's mare, and went over to the East Gallatin to see about a school there.

Sunday, July 22, 1894

Stayed at Mr. Yates last night. Stopped to see Mr. Davis, another member of the board. They seemed perfectly willing to give me $60 but I do not know whether they will give $65 or not. They are to let me know soon.

The trains are running now but I do not know whether they are running regular or whether they have come to an agreement concerning the strike.

⌘ ⌘ ⌘

A GROWING ENTHUSIASM:
Attacking a Fossil Skull

More and more bones and fossils were found in the Madison Valley. My father took side trips into other nearby areas and, after returning to the local hunting grounds that supposedly had been stripped of all fossil remains, found more and better material. He visited Big Round Top again and again and each time he made discoveries. He discovered Little Round Top and so many other round tops that naming was abandoned and they were numbered.

His needs increased. He wanted a horse and buggy of his own so that he might explore other areas of the state to study geology and collect fossils. Identifying the fossils and establishing the geological age became of primary importance for he believed some were new to science. He needed scientific books, which were ordered as fast as funds could be spared. There was little money, however, as practically everything earned was going toward reducing his debts accumulated in college, or for his folks.

My father applied for the school in the Gallatin Valley near Belgrade, not because it was in a better location geologically, but because it offered a better salary and a faster means of reducing his debts. School teaching had become a means to an end. Geology and paleontology were fast becoming his primary interests. With the money he received from teaching he could live, repay his debts and buy books so that he could delve more deeply into his favorite subjects without the torture of poverty. His enthusiasm was suddenly crushed when he received the news that his mother was ill and needed an operation.

August 8, 1894

I was shocked to learn that Mother has a tumor and it is a very difficult place to operate but it must be done. Her health has been poor all summer, something unusual for her. I had always thought of her as living to a good old age and this is wholly undreamed of. It will cost a good deal of money.

Getting a position in the Belgrade school seemed doubtful for awhile. He feared he had asked too much but was reassured when he finally received notice that he would be employed at $65.00 per month. He boarded at Mrs. Davis's for $3.00 per week.

Belgrade some eleven miles northwest of Bozeman and almost in the center of the rich and scenic Gallatin Valley was not nearly as suitably located from a fossil collecting standpoint as was the little school on the Madison. Fossil hunting was not completely abandoned, however. In the Bridger Range to the north were the nearest exposures where fossils might be found. Although nearly ten miles distant my father managed to explore them either on foot or with a borrowed horse. These formations were the old limestones and the fossils were principally ancient sea life. Valuable collections were made of these invertebrates.

One of the few sports my father cared for was duck and game bird hunting. He did not indulge in them to any great extent, however, for his work came first. He never cared for big game hunting. At Belgrade it seems he hunted more than usual.

Saturday, November 10, 1894

Hiram Ballard and I took a team and started for the mountains hunting. . . Drove in the canyon about as far as we could go with the team and ate a lunch and then started hunting. Went along the steep ridge on the north side of the canyon. It is high and covered with broken rocks - the gneiss and Archean formations which appear on the west side of the Bridger Range.

Shot a couple of grouse on the north side of the ridge in the pine timber. Shot 4 more and Hiram shot one making 7 in all. I got into camp about dusk and Hiram came in later. Had a camp fire and cooked some grouse and potatoes.

Sunday, November 11, 1894

Had a good nights rest out under the open sky. There was a little breeze but we were in a sheltered spot and it did not strike us much. We could hear the roaring of the mountain stream all night and the sighing of the breeze in the pines and firs. We could see the moon and stars above.

Went hunting again north and N. E. of camp. I killed 7 grouse and it almost killed me to get them home. These mountains are awful to climb especially with a load.

Got started home a little before sundown. Had a pleasant ride home only the road was rocky part of the way.

Wednesday, December 12, 1894

Went down to the little pond or biou (bayou?) west of here to shoot ducks.

I waited for some time hoping that I would get a shot at some. At about dusk they began to come in. Two came and I shot the drake, the duck came flying over and she fell at my very feet. Shot a good many times but I only got one more. Had considerable fun; they will light at ones feet in the evening.

Everything at hand was being read that would shed any light on the identity of the fossils.

December 13, 1894

Arose early and read from Hayden's Geol. Rep. 1870. It is interesting to me. Was reading about the Loup Fork and White River beds. There seems little doubt that my fossils are most nearly like those of the Loup Fork Pliocene. I think I have rhinoceros, mastodon or elephant, one or two horses, camel or precamelus. I think next month I must have Leidy's Extinct Mammals of Dakota and Nebraska.

Received a letter from Prof. Foster. He says they are preparing to start the University next Sept.

Had it not been for teaching in a small crowded school house, much in need of repair, Father's winter might have been quite pleasant. The enrollment started with twenty but soon increased to 33 and later around 65. In spite of the fact that the school began satisfactorily it soon became a source of irritation and frustration. Some of the students were discipline problems and the extremely crowded room did not help matters. Perhaps another factor that contributed to the school's problems was the teacher's fading interest in the students and teaching. Instead he was experiencing a growing enthusiasm for the study of the rocks and the fossils they contained. His interest in the latter was increasing at a rapid pace in this fascinating land so favorable for their study. He completed the school term, however, but not without considerable grief to the students, to members of the school board, and to himself. By March 1, 1895 the day the school term ended, he was glad to return to the little school in the Madison Valley and teach for $45.00 per month but unfortunately he had to board himself this time.

On February 17th Father purchased a horse which he had so long wanted. The agreed price was $35.00, $10.00 cash, $15.00 to be paid in three weeks and the balance the first of June. March 8th he received word from Mr. Kay, the man from whom the horse was purchased, that he had sold another man's horse. Upon returning the horse he was given a different horse, the price being reduced $5.00. The new horse, which he named Bonepart, was not nearly as good. On March 14th he purchased a buggy for $25.00, paying $13.00 in cash, the balance was to be paid May 1st.

On March 16th Father packed and started for the Madison with a newly-purchased horse and buggy which he had obtained for $55.00. Upon reaching the Madison Valley little time elapsed before he began collecting fossils again. He now began to spend more time in removing the bones from the rock and began employing new techniques to preserve them without injuring the precious material.

Sunday, April 21, 1895

Today I went up a run farther north. Went nearly to the top of a kind of pass then went to a bluff to the north . . . I found just above a little clay exposure some small pieces of bone and teeth. I picked them up and concluded they were rhinoceros teeth. Dug a little more and found there was more bone there so went over to where my horse was and got a pick and ate my dinner and fed my horse and began to dig. I dug carefully and found that there was the greater part of the skull with the upper teeth. Then I found both of the lower jaws both with teeth... I dared not go further as they are very frail. I fear I cannot get them out whole but am going to work carefully and try. If I can in anyway get them out and preserve their shape I intend to do it if it takes a week. They will be a valuable specimen. I drew the upper teeth top view and intend to do more drawing before I expose them further.

Sunday, April 28, 1895

Today Mr. Hutchison and I went down east of Crowley's where I found the rhinoceros. We used various implements to apply paraffine and plaster of Paris. This was my first experience at mending bones, but we did well. The lower jaw, one part of it, we got out complete, only breaking off a few pieces from one or two teeth, lifted it out and then put it in a box with plaster paris. We then attacked the skull and did not do so well. It broke in spite of all we could do but we got out a full set of 7 teeth and quite a portion of the side of the skull.

On May 8th my father received a letter which was of great help and encouragement:

"Received a letter from Professor Scott of Princeton which was of course very interesting. He thinks I ought to go east with my collection as it is almost impossible to do satisfactory work where the specimens and literature are wanting.

> If I had the money I would go next winter but there is little prospect of it now.
> He says that a thin solution of gum Arabic is excellent to stiffen crumbling bones and that a good plan is to put narrow strips of muslin covered with flour paste as soon as the bones are uncovered and remove them and cement them after they are taken home. Scott seems to be accommodating and evidently is willing to do what he can for a fellow. I think I will soon give him more data, send him drawings etc. He is a good man to keep on the right side of I think."

Later in May he discovered his first fossil of a carnivorous animal.

> Monday, May 13, 1895
> Today, before school and after, I worked on the teeth of the carnivorous animal I found Saturday. It was broken into scores of pieces and it is a very interesting puzzle, getting it together, but I am progressing nicely. The teeth I have gotten nearly together I think. I can nearly reconstruct one side. The skull must be there, a good share of it, but it is very much broken. I want to make all I can of it as it is the only carnivorous skull I have.

After completing the spring term of school, my father stayed in the Madison Valley and contracted to teach a fall term in an adjacent district. Mr. Darlington, who worked for him in later years, was a member of the school board. He was to receive $50.00 per month and board himself. In August he lost his horse.

> Saturday, August 17, 1895
> Bonepart died today. He got sick and was suffering great pain and I asked Mr. Tice to kill him. He never has been all right since I got him, I think.

Father rented a room from Mr. Bishop and moved in September 8th just before the fall term began on September 10th. He reviewed his financial circumstances October 21.

"It may be interesting at some future time to know my circumstances at present. I am getting toward the shore of my small sea of debts. I owe probably about $75 or $80 now but I am actually getting ragged and have not a real good suit of clothes or overcoat to cover my back. I cannot at present buy clothes either as I have not money and see no assurance of having any. After my debts are paid I shall not have enough to keep me through the winter. My school is for only 3 months. But the worst is my folks are worse off than I am and I want to help them so much but it is not in my power at present. I need books very much indeed but I cannot buy them at present. I need many things but I must continue to want. This is the dark side. The bright side is I am doing considerable in geology and my success helps me to brave these ills.

I went up to Bozeman last Friday evening to get a horse. Stopped at Professor Foster's as usually I do. He concluded to let me take Prince to keep through the winter."

After not visited Big Round Top for some time Father decided to stroll up there, although not expecting to find much after collecting so many specimens from its slopes.

January 29, 1896

"Today has been one of my most successful ones if not the most. I first found two turtles in the same block of fallen clay below the cliff in Big Round Top, one just behind the other, and I think different species. Then I found the mandible remains of a horse. I then went above the cliffs on the slope back of Big Round Top and found a little bone sticking out of the clay. This may be a little camel. . . . I found more bones and finally the whole lower jaw and cervical vertebra, at least 3 or 4. The jaw is turned wrong side up and I believe the skull is there. I dug around it leaving a block of frozen clay containing the bones. Expect to get it in the morning. I expect to bring it home and get the bones out in good shape if it takes a week or two. I hope and believe it can be done."

Father taught the spring school term in the same little schoolhouse he had started teaching in two years previously. Sometime during the term he made a decision to leave the Madison Valley as soon as the school was completed to explore new fields.

⌘ ⌘ ⌘

THE RUBY VALLEY:
New Scenery

My father now seemed to consider the best plan for his future in paleontology was to remain in the Lower Madison Valley, continue his collecting, and concentrate on the study of his fossils, with the idea of taking them east to determine their identity as soon as possible. This seemed to be the dictate of his better judgment but the temptation to explore some of the other valleys, beyond the mountains, was too strong to resist.

There were several areas he wished to visit but nearest at hand was the upper Madison Valley, known now simply as the Madison Valley. From there it would be easy to cross the mountains to the Ruby Valley, visiting Virginia City on the way. From the Ruby Valley the Beaverhead Valley and Dillon would be easily accessible.

Prince, the horse Professor Foster had loaned him, was returned in April at Foster's request. Soon after that he purchased a horse, which he named Pete, that was to remain with him for several years. Once more he was able to travel by horse and buggy.

Although he had little money, and no school in sight for the fall, he was determined to leave as soon as school was out. Before leaving he wrote a detailed description of the valley and its geology. He also carefully boxed his specimens and stored them in Bozeman. By July 8th he was ready to depart from the place where he had spent so many happy hours exploring in the hills.

>Wednesday, July 8, 1896
>   Left the lower Madison Valley, where I have been so long, yesterday.
>When the Middle Madison came into view I was delighted. On the east and west were high mountains. About in the

center of the valley the river flows through numerous channels, each lined with trees. To the east are low cliffs forming the eastern boundary of the terrace which, smooth as a floor, slopes gradually upward against the high abrupt mountains. The terrace is evidently lake bed deposit, but I have not examined them yet.

Thursday, July 9, 1896
   Did not get started early on account of looking for Pete. He is not good to stay with a fellow. Stopped at Ennis for some time. Got my buggy fixed a little. Had talks with several and learned considerable about the country. Came up to Cedar Creek Canyon and camped. Mr. Thornton lives just below the mouth of the canyon. Met a bear hunter just coming out of the mountains with a bear skin. His name was Wheary, or something like that.

Saturday, July 11, 1896
   I am still camped in Cedar Creek canyon. Yesterday I started to ride over the divide to Jack Creek, but a high mountain to my right tempted me to climb it as I wished to get a view of the surrounding country. It looked as though it would be impossible to climb the highest point, but I had heard that it could be climbed so I went on. We had to cross a long slope of thin, loose blocks of rock and Pete did not enjoy traveling on them very much, especially coming down. I ascended to the highest peak and got a fair view. On the north the high rugged peaks shut off the Gallatin and Madison Valleys. To the east, southeast and south was a wilderness of mountains and wooded canyons. To the west I could see over into the Beaverhead country and beyond with its white peaks gleaming in the hazy distance, what I suppose to be the main divide of the Rocky Mts. Descending the mountain I found a quite good exposure of cretaceous slates. Found a species of shell which I have not found before, but best of all I found at last what I long have sought,

Cretaceous sandstone with impressions of leaves. I felt well paid for my hard work.

I am having a grand time and am happy. Seem to be just in my element.

Sunday, July 12, 1896

Yesterday afternoon I packed up, broke camp and drove over to Jack Creek Canyon. Had an accident on the road when I went to drive through a little irrigation ditch, split a thill (wagon shaft, DDI) and turned a wheel wrong side out. It came near upsetting the buggy. Pete fell down broadside. I was lucky to get out as well as I did. Stopped at Mr. Bull's, got some oats and had quite a talk with him.

Am camped now up above Jack Creek Canyon perhaps 2 1/2 or 3 miles above Mr. Hutton's house. . . . Last evening I walked down with him and got some milk. He told me about a waterspout which flooded the canyon and changed the course of the stream. I thought we were going to have another like it today. The clouds below the mouth of the canyon were dark and moved rapidly. Then they piled up in large masses on the peaks to the north and the lower part was in a commotion. I took my bedding over to a cabin but it did not get wet through the tent. It rained hard and the wind blew considerable.

I went fishing this forenoon and caught 10 trout. One was a nice big one a foot long. I had some for dinner and they tasted good.

Sunday, July 19, 1896

Thursday, Fred and Perry Hutton and I went up to the V-shaped mountain where I got the fossil leaves. It was a long climb. We left the horses further down than I left Pete before and hobbled them. I got some new specimens of leaves. When we returned the horses were not there and we could not find them so we had to walk home and it was a long hard walk. I went down to Hutton's with the boys and stayed all night. Had a pleasant evening. Miss Mary Bull is there; she is also interested in fossils. Mr. and Mrs. Hutton are also nice company. Fred is only 11 years and is in the 9th grade in the Bozeman High School.

Yesterday I packed up and drove down to Ennis, got some repairing done. I left a box of mostly fossil leaves at Ennis. There was nearly 100 lbs.

Friday, July 24, 1896

Monday I stayed on Wigwam Creek and found a few crinoids.

Tuesday morning I started for Virginia City on the Ruby Valley, went by Mr. Thompson's to see some bones of an animal which were found on the "Gravely Range." He thought it was a mastodon but it was a sea monster of some kind... probably came from the Cretaceous.

I reached the Ruby Valley a little before night. I stopped in Alder Gulch a little this side of Virginia City. A Chinaman was separating gold with a rocker. He said he did not get much. I went further down where there was quite a large machine at work and had a long talk with a young man who was running it. He had an elephant's tooth which was breaking to pieces. He said I could have it if it would be of any use to me and I gladly accepted it.

I stopped that night at Dr. Amsden's. He is a horse man. I met Mr. Raymond also, another man who raises horses.

Wednesday afternoon I went to see Mrs. Whitcomb, the Co. Supt. and had a very pleasant visit with her. There are several schools not taken. Thursday morning I started up the Ruby. I stopped to see Wm. Taylor, a member of the school board in the district where Dr. A. lives. I went through a small canyon and then came to a broader valley with lake bed deposits. I examined them but found no fossils but small fragments.

Sunday, July 26, 1896

Night before last I stopped with Mr. Raymond on his horse ranch. He has a nice ranch, 2 large barns and a good house.

Two or three rooms in his house have nice carpets and the things are kept in nice shape. It looks as though there ought to be a woman there but there isn't. It is much the same at Dr. Amsden's. Dr. A. has fine heads of deer mounted and his two best rooms are decorated with them.

The Raymond ranch was on the Sweetwater, a tributary of the Ruby River, some twenty miles south of Virginia City. My father continued his journey up the Sweetwater in a southerly direction, crossing over into Blacktail Creek, then down the creek in a westerly direction to the vicinity of Dillon. July 28th he was camped about seven miles east of Dillon and was searching for the place Mr. Steers, the man he had met in Helena two years previously, had found the bones of an extinct horse. He was looking for a school and had contacted two members of boards. He said it made him sick to think of teaching a small school another winter and he was going to make an effort to get out of the business before long.

> Tuesday, August 4, 1896
> When I was camped far away from anyone I was taken sick and have not written in my journal since. I went up on the mountain and searched the region with a field glass. I saw one place where a white substance was exposed. It proved to be the place and it made me quite happy. . . I intended to go to the place the next day but my head began to ache before I got home and I awoke with a head ache the next morning. Thought I would have to see a Dr. but could not well afford it. But I felt better when I got in town, got some salts, drove back to the bluffs east of town and camped, took a lot of salts and in the morning I felt better. I thought what an awful thing to be sick among strangers. But this was not all my troubles for I received news from home that mother was very bad off again. I had been in hopes I would see her again but it is very doubtful. I would be tempted to sell my specimens if I would get a chance.

Thursday, August 6, 1896

After learning of Mother's sickness I decided to settle down and get a place to work if I could and be where I could get my mail. I began to inquire Sat. and continued it the next day while driving in the direction of the Ruby. Stopped at Walden's horse ranch and took dinner Sunday, struck Sheridan a little before noon Monday. Heard of a place to work on the Thompson horse ranch. Came down and got a job at $1.25 per day, began work in the afternoon.

Wednesday, August 26, 1896

This is the fourth week I have been here. Have been at work in the hay fields nearly all the time. Was nearly sick when I began work but improved and got very hearty yet the work tired me. Yesterday I was sick in the afternoon and quit work.

I have engaged the school south of Laurin in the district in which Dr. Amsden lives. I get $60.00 per month and expect to keep house myself.

Thursday, September 17, 1896

Have been teaching nearly two weeks now. The attendance is improving and all are punctual. I have 35 pupils enrolled. The first week I had to leave my handy quarters by the school house as Mr. Conry traded places with Mr. Hinch. I am nearly two miles from the school house. I drive to school so it does not seem far. I like the children here very much. Have not been around visiting the people as much as I intended to.

At this time my father had made a complete circle of the Ruby Mountains that form the western side of the Ruby Valley. He noted that he had traced the lake beds all the way and he pictured the country in prehistoric times as being a huge lake with the Ruby Mountains a huge island within it. He did not gather any great amount of unusual fossils on this trip but the geology of the country was a revelation to him. He was soon to start making plans for a similar trip the following summer. His greatest worry now was his mother. In October he received word of her declining health and soon afterward her death.

> Thursday, October 15, 1896
> We are having splendid weather now. I think I could enjoy myself very much were it not for Mother; she is failing.

> Friday, October 16, 1896
> Since writing last I have received the new of mother's death. I did not know but I would die myself. My heart acted very badly, the worst it has ever done. It would beat in all sorts of ways and then it would not seem to beat at all. . . It is evidently growing worse yet for a long time last summer it did not seem to bother at all. My time may be very short. . . I think I'd rather die of heart failure than any other way. I hope it will take me without a moments warning.
> It is terrible to lose a mother. . . I have been away from home so long that I will not miss her like they will at home but there is none left to love me with a mother's love now. There will never be another so faithful and true. None like her can charm away pain.

⌘ ⌘ ⌘

16

EXPEDITION OF 1897:
The Yellowstone Park

Teaching in the Ruby Valley, to my father's surprise, was much more pleasant than his previous teaching in Montana had been. The school was considerably larger than the schools in the Lower Madison Valley, but not so large and crowded as at Belgrade. He had found a nice community and was beginning to enjoy it. There was one thing, however, that he thought might cause trouble. He feared he might become infatuated with some of the nice-looking older girls in the school and would have to leave. "In fact", he said, "there was already one that he thought a great deal of."

Of greatest importance at this time, however, was the contemplated paleontological expedition for the coming summer. The possibility of having the companionship of an interested school teacher, whom he had met at the teacher's institute in Virginia City, stimulated his preparations.

> Sunday, November 1, 1896
>
> I attended Teacher's Institute all last week beginning Monday. I gave a lesson each day on the study of rocks and much interest was manifested by many of the teachers and people from town. I had never given lessons in an institute before and did as well or even better than I expected. I had a display of rocks and fossils of the region. My work was a good deal of trouble but I felt paid, as I am sure many got some new ideas and then they were very appreciative. Among those who showed special interest was a Mr. Chas R. Parry of Red Bluff.

Tuesday, December 1, 1896

Friday I could hardly get to school without freezing and Eddie froze one ear badly. It has been −30° or more. I finished three months of my school yesterday. They say I am giving good satisfaction and most all are pleased.

I engaged the school for three months at first but I may stay the other six. A small school like that on the Madison did not bring out all that was in me and I felt somewhat ashamed for teaching such a small school but I feel different here.

Financially I am improving slowly. I have all my debts paid except $28.00 and a few little ones but I must have a few books and some clothes.

Mr. Parry, the teacher whom I met at the institute, and who took such an interest in geology was here to see me last week. We are planning to go out together next summer. He has a horse and camping outfit; we will have to get a wagon and harness. We ought to have a camera too. We want to start about June and stay until October. We want to visit the Big Hole Frying Pan Basin, the Upper Red Rock, the lake beds of the Upper Blacktail, The Gravely Range, Upper Madison, Yellowstone Park, and visit the Lower Jefferson. I think next summer we shall be able to partly pay expenses by selling specimens.

Wednesday, December 23, 1896

My school takes a good share of my time and I sometimes regret that I have so little time for geological work. I am preparing an article for the Madisonian on the geology of the region. Though it is for a country newspaper I want to do my best. Mr. Cheely, the editor, invited me to write the article.

Sunday, December 27, 1896

Last evening we had our second meeting of the Literary Society. It was a great improvement on the first and I hope we will have a good society.

Father elected to continue with the school for the remainder of the term. On January 8, 1897 he was preparing a large chart on the geology of the Ruby Valley and surrounding country to be used with a lecture he was going to give. The lecture was later given in Virginia City and a collection was taken up but it did not pay for the expense of the chart. At the end of the entry the names of the students taught that term are listed and among them was Pearl Goetschius (my mother) and her younger sister Edith. Their father was John Goetschius whose ranch was about a mile and one half from the school house. In February Father and Mr. Goetschius went fossil hunting.

> Friday, February 26, 1897
> I had not got around to mention my finds when I wrote last. Sunday when Mr. Goetschius and I went into the mountains we found many fine specimens. In fact I think it is the best place for subcarboniferous specimens I have found yet. . . . It is so handy too. It is right at the edge of the Ruby Mts.

Father and Mr. Parry started their trip in June as planned.

> Tuesday, June 8, 1897
> We, Mr. Parry and I, are now camped above the lower canyon of the Ruby. We have started our summer's trip. We started a little before 1 o'clock yesterday afternoon.
> We have been looking up the Miocene Lake Bed deposits a little this afternoon. find no fossils except some petrified wood. . . .

> Upper Ruby Valley, June 16, 1897
> We are camped now near a little rivulet on the west side of the road near Mr. Garrison's. There are about 5 families in the valley. I do not think they raise any grain here but they raise plenty of feed for horses and cattle.
> It began to rain last evening and rained all night and most of the day.

Saturday, June 19, 1897

We broke camp near Mr. Garrison's today. We wanted to get over into the Red Rock country as we were nearly out of provisions. This country is excellent for cattle in the summer. The feed is good now and our horses get plenty to eat. We have had about 3 days of pretty bad weather, cold rain and snow a little higher up. We suffered some with the cold and our horses must have suffered much a couple of nights.

We are encamped at the Three Forks of the Ruby this evening. Have been following the river, east of the Snow Crest Range. Its highest peaks and crests are very prominent. There is one mountain and canyon that I especially wish to visit and think I would soon if it were not for the snow. We have not seen the range to our east, or its highest peak, Black Butte, which stands up so prominently today.

Tuesday, June 22, 1897

Have been encamped since Saturday afternoon at the Three Forks of the Ruby.

There is a road going to Red Rock Creek but it must be nearly 30 miles that there are no inhabited houses. Mosquitoes are plentiful but they do not bother at night, it is so cold.

I am enjoying the trip very well especially since I have felt that I was accomplishing something. It is hard writing now for I have no table and it is hot and the flies and mosquitoes bother but it has to be done.

Yesterday and day before I went over to Devonian Mountain. I call it Devon Mt. because it is composed mostly of Devonian rock and I need to mention it often. It was a revelation to me. At a distance I could not guess with any certainty what the strata was. When I got to the flank of the mountain I was a little surprised to find Devonian fossils in the limestone. Mr. Parry went down the Mt. and I went up. I soon came to rock containing large brachiopods. They are the largest I have ever seen. I was very much elated over my find.

I am sitting on the brow of a hill where the breeze can cool me and blow away the insects. . . . The sky is a beautiful blue and it just struck me that there is no blue like that of the sky. It cannot be described only it is sky blue. There are white clouds broken in many fragments and scattered beneath the blue. All the clouds are changing every minute.
Evening in tent.

Have spent the day very practically in camp-- writing, studying fossils, especially brachiopods, and getting my fossils in shape. Among my fossils, as near as I can determine, at least 12 are different. It is a rich hunting ground.

I forgot to mention seeing a flock of Big Horn or Mountain Sheep yesterday. They were across on another mountain and I looked at them through my glass. There were 9 or 10. They moved along leisurely. Their horns are enormous.

Wednesday, June 24, 1897

I arose at about 8 o'clock this morning. Went fishing a little while, caught 5 trout, small ones. We did not get started from our camp at the Three Forks until about 1:30 P.M. The roads are not good at least they are quite bad a part of the way. There are a good many mud holes. One place we cut a road through the willows as it was very bad. We drove 10 or 12 miles. We crossed the divide at 5 o'clock. The Snow Crest Range continues toward the S.W. with about the same general appearance.

Friday, June 25, 1897

Camped north of Red Rock Lake
Yesterday we started at about 10:30.

We came to a light colored deposit which I thought at first was lake bed deposits but found it to be Cretaceous. It is the same formation I examined at Warm Spring Creek hoping to find some vertebrate remains. Went farther south to a small exposure and found vertebrate remains. This was quite a discovery to me as I had never found any saurian remains and I was so anxious to find the stratum in which they occur. I found a tooth and several incomplete vertebrae. The bones were of a small saurian of some kind.

We stopped at Mr. Metzals north of the lake and got flour, butter, eggs, meat, potatoes and some dried fruit. We were extremely glad to get these things as we were almost out of provisions. Their ranch is the only one I know of on the N side of the lake.

Friday, July 1, 1897

My time has been so full or occupied lately by the new sights and my desire to see all I could of them that I have written very little. . . am now in the Upper Geyser Basin of Yellowstone Park.

Saturday, the day after my last writing, we stopped to inquire the way of a gentleman in a house by quite a large creek. He thought we would have no trouble going to the park as the flies had not got very bad. He showed me a book by an amateur explorer, named Brown I think. He makes the true source of the Missouri to be at the head of the Red Rock River in the Rocky Mountain divide in Montana perhaps 10 or 15 miles east of Red Rock Lake. We were much annoyed by the mosquitoes it being the worst night we had spent. I forgot to mention the night before we had a rocking thunderstorm and hail pelted on the tent until it seemed it would come through. The horses, which had been very good about wandering off, were found perhaps 1 1/2 miles back. The next day we had good roads except a little spot N. of Henry's Lake where Pilot got swamped in the mud and we got stuck and had to unload and hitch the horses on the hind end of the wagon and pull it out backward.

We stopped at Mr. Sherwood's store and got provisions. The next night we camped perhaps 5 miles from the store. The mosquitoes were terrible but I managed to sleep pretty good as I slept out doors with mosquito netting over my face.

That evening a queer circumstance occurred and it caused considerable fun. I heard Pete's bell, which I had just bought for him, but could not see him. After I returned to camp we heard the bell rattle as though the horses were traveling real fast. Finally we started down that way and there seemed to be a man on a horse driving them away. I was much surprised and could not understand it but of course thought "horse thief." He had a red jacket and as he was persistently driving them away how I longed for a six shooter. When I got a little nearer I said, "What's the matter." The horseman immediately turned and rode right toward me. I discovered it was a woman. She explained the matter. A gentleman staying with them had lost a couple of horses marked with an O. She saw the O on Pete and thought they were the horses. She asked us to be sure and stop the next day at the next ranch where she lived. We had a hearty laugh after she had gone.

The mosquitoes were so bad next morning that we started early without breakfast.

The flowers along the road are of great interest to the botanist and I wished I could find the names, but we had not the time. I cannot do what I want in geology and do much in botany. Some of the flowers are beautiful.

At about 8 or 9 o'clock we arrived at the residence of Mr. Durelle. It was Mrs. Durelle who tried to drive our horses away. She is a very friendly woman. The house seems in keeping with the surroundings and situation. The main part is square and built of logs. Inside one enters a large room, carpeted. On the floor are bear skin rugs with the heads on. On the banister which leads upstairs are other large ones. There are elk heads around the room and many trophies of the hunt and chase. There is a stand on which are books and a small book case on the wall which is full. Mrs. Durelle brought out a lot of skins of otter, mink, martin and coyote and showed us. The house impressed me very much and I thought how fitting for such a wild romantic region. Mr. Durelle is a hunter and evidently loves the sport. They entertain travelers I think and it is kind of a stopping place for people going to the park.

After leaving this place we went through nearly level pine country. All among the pine woods are open spots smooth and beautiful. This kind of country extends for miles and the park begins in these forests. The only thing that tells where the park begins is, "Park Entrance" put on a tree and near it a copy of the Rules and Regulations. (this is what is now West Yellowstone, GED).

The park seemed quite dull and uninteresting to me until we entered the Lower Geyser Basin. The rock is all, or nearly all, dull forbidding looking lava. We are now in the Upper Geyser Basin. Primrose Lake and Turquoise Pool are very pretty and well described in Haynes Guide Book.

I had an interesting talk with a very pleasant gentleman and his wife from the Atlantic Coast. He seemed interested in my geological research and I told him a good deal about it. I also made the acquaintance of Dr. W. Dodrick from Hong Kong, China who is an astronomer. I had a very interesting talk with him while waiting for the fountain to play.

Old Faithful is the only geyser that has come up to my expectations, or the only one which, when I have seen it, has not given me the feeling of the lack of power.

July 11, 1897

I am sitting now at the foot of a pine tree on the edge of a bank on the west side of Yellowstone Lake. The view from here is pretty. Just below is the shore of the lake with a narrow dark sandy beach with sand and gravel arranged in more or less regular lines parallel to the shore.

The water of the lake is clear and the beach clean. Back of me are pine trees with almost no brush or fallen timber and there is much grass and quite a good many wild flowers. To the north there is a grassy lawn back of which are dark pines. Back of that is a high ridge covered with pines. Beyond that is a mountain slope that looks as white as snow.

It looks like the deposit of some enormous geyser. Beyond it and toward the north is a high ridge covered with dense pines. To the right and beyond the ridge is a rounded peak, then there is a pretty 3-lobed ridge, then a lonesome one and the most interesting one of all, a great human face as if lying in state with the face turned upward to the sky. There is something in the face hard to express but it has made me feel as nothing in the park has done. I can now understand Hawthorne's Great Stone Face and how it could impress one so profoundly.

Caught 24 fish yesterday. Have had good sport fishing and feel quite satisfied. Have salted nearly a keg full for future use. They are nice large fish. It has frozen the three nights we have been at the lake but is hot at midday.

Thursday, July 22, 1897

I have had considerable experience since I last wrote but have not had time to write. The 13th we broke camp and stopped at a large house near the hotel and got some provisions. We camped about 3 miles from the Grand Canyon Hotel.

In the morning we were surprised to learn that the soldiers had ordered us to shoot our dog. I do not like to dwell on the affair as it caused us so much trouble and anxiety and anger and I will remember it as long as I live probably. I have felt at least if someone had ordered me to do the same by him I would not shed a tear. The order was so dastardly, mean, unwarranted and contemptible and he showed the spirit of some old eastern tyrant. But it is a pleasure to know there are such men as Mr. Walker, manager of the hotel, who saved the life of the dog when all else seemed hopeless. He is a man with a heart and I was surprised to find a man like him.

The next day after this affair we started for the Mammoth Hot Springs. Arrived there the next morning early. I went to the P.O. and got my mail. Found by the paper that I had been elected to the principalship of the Virginia City Schools.

The next day we went through Norris and east 4 or 5 miles on the road to the Grand Canyon. Camped and rode horse back to the canyon. I looked at the canyon with great satisfaction, also the falls. At about sundown we went to the soldiers' camp, had a talk with the boys and got the dog. It seemed I never saw a brute or human being so wild with delight. He stood and trembled awhile hardly knowing what to do, then the wildest demonstrations of delight, and when we started back on horse back he went pretty near crazy. Poor dog, we had all our trouble and anxiety for nothing and his pleasure was short.

The next night we camped at the mouth of Gibbon Fork and the next outside the park at Durelle's. Mr. Parry did not feel well and I had a lot of washing and straightening up to do so we did not travel the next day. The next day we went only 6 or 7 miles and camped near a creek as I wished to go to a mountain where there seemed to be such an excellent exposure of rock. I rode Pete and had the dog with me.

The dog got after a porcupine and got hundreds of the spines in his mouth and all over and inside. He was almost crazy and I thought there was no hope for him. I tried to get them out but gave up in despair, and killed the poor brute to get him out of his misery. I had a rather dangerous experience too. I wore my old hobnails -- slipped on a steep slope and went sliding down at a rapid rate without power to stop myself, over the bare rocks and a low precipice, brought myself to a stop below. I was a little anxious about my life but escaped with no more injury than bruising my hands.

Monday, July 26, 1897 - Red Rock Valley near Lima

On the road from Red Rock Lake to Lima I did not stop to examine the rocks or look for fossils. After leaving the valley of the Red Rock we ascended a quite long hill then went down into the valley in which Monida is located. It is a little valley about a half mile from the Idaho line. I found 6 or 7 letters at Monida, which have been delayed by somebody's mistake and I fear it will cost my position at Virginia, and it is too late to get a good school now I fear.

July 27, 1897 - Sage Creek about 8 miles north of Lima.

Have been very successful. I found the milk tooth of an anchitherium (an ancestral horse of the Oligocene) and part of a rhinoceros with two perfect teeth, a couple of vertebra of a turtle and the limb bone lined with quartz crystals, very peculiar it seemed to me. I feel our trip is not going to be a failure.

Wednesday August 4, 1897

Camped on a branch of Sage Creek. We wanted some supplies and I had seen some fine exposures of lake beds so came this way instead of going across to Blacktail. Came by one house where they were haying, stopped to see if I could get some butter, lard etc. I got what we needed, made my business known, found a young lady, Miss Ada Talbot, who was very interested in collecting fossils. She got me very much interested, showed me a few specimens and we had a very interesting talk. I concluded to find a camp as soon as possible. Found a splendid place only a little way from there – a nice little stream, a delightful place to pitch a tent among the willows and plenty of wood and good feed.

Monday we followed the road up to Mr. Freeman's where it ends, and went over the hills toward Dillon.

By a little stream I killed a couple of young grouse, knocking them over with stones. Came to where there was a herder with a band of sheep, about 2800. He seemed quite an interesting young fellow. He told us to go to camp and cook us some food and help ourselves. We made some bread and cooked our grouse, then Mr. Walker came and we had supper.

August 5, 1897

Yesterday a little after noon we started for the Blacktail. I did not think to inquire about the road. We soon mistrusted we were not on the right road but did not like to turn back so when all hope was gone we turned off the road to see if we could find the right one. It was rough and rocky and muddy and steep and sidling and brushy, but we made a go of it. We found a good many crystals that we would not have found had we not got lost. Priest Canyon is terrible rocky and it was very unpleasant descending it.

August 10, 1897

Thursday we broke camp and about noon went east up the Blacktail. We came to a place where there were three branches. We found an excellent camping place.

We took a look at some deposits. It is a light grey color principally clay. I found one fossil--part of a jaw of a small rodent with four broken molars and a broken canine.

The next day we got on our horses and rode to a large exposure on the left fork. The rock there is very different. It weathers in peculiar shapes, pillars and other architectural forms. It is the finest thing of the kind I have seen.

Frying Pan Basin, August 17, 1897

Came here Saturday afternoon. We broke camp on the Blacktail on account of trouble with my tooth. I wanted to investigate there a little more but could not stand the pain. Did not get a good nights rest for over a week. The nearest I came to it was when I got nearly drunk and slept 4 or 5 hours perhaps.

We are now camped on a side hill near a cabin by a ravine in which is a spring. We are, I should judge, a little northwest of Dillon.

August 20, 1897 - Camped on the Big Hole River.

After leaving Frying Pan Basin we went to Dillon. I got some boxes, drove out east of town and packed my specimens and left them in Mr. Watson's store. The load is lightened several hundred pounds.

It is about as dry as a desert around Dillon and almost no feed. We had a hot ride today. We drove down to the Big Hole. We are camped on an island and there is a little grass for the horses and plenty of water and wood.

Father and his friend continued their journey on down the Big Hole and Jefferson Rivers to Three Forks and from there into Bozeman. Father wanted to see if his specimens that he had stored in Bozeman were all right. After leaving Bozeman they visited the Hoelleins and other friends on the Lower Madison. He said they had a jolly time and didn't know of a place he would rather visit. After a day or two on the Madison he proceeded to Virginia City, where the school board held the position open in the high school for him. He started teaching Monday, September 13th. The summer's trip was over but it was not to be the last of his fossil hunting expeditions. Before the snow had disappeared the next spring plans were laid for another trip the following summer.

⌘ ⌘ ⌘

EXPEDITION OF 1898:
The Bitterroot Valley

As principal of the high school in Virginia City, my father found little time for anything but school work. Latin and German were new subjects to him as a teacher and kept him well occupied. He did not completely forget his fossils or the study of geology, however. He carried on a fruitful correspondence with the United States Geological Survey to determine the identity of his fossils, especially the invertebrates. In January, 1898 he sent over one hundred specimens to Dr. Wolcott, head of the survey.

Another interest that he now acquired was the geology of the mining district in which he was living. Alder Gulch, in which Virginia City, Montana is located, was one of the richest gold placer deposits known. Although most of the gold had been extracted from the rich gravels by 1898 (except that taken out by the dredge boats which were to follow) the source of the gold was still a matter of speculation. The lode mines that had been discovered in the gulch did not seem nearly rich enough to justify the fabulous wealth the grave had yielded. Money was still being spent to locate a rich gold vein that most people in Montana believed existed somewhere at the head of the gulch. Father's study of the source of gold in Alder Gulch was later presented as a published article in a mining magazine.

The Ruby Valley was about nine miles distant. He occasionally visited the acquaintances he had made while teaching there the year before, among whom the Goetschius family were visited somewhat more often than the rest. The death of the five year old son of John Goetschius in September saddened him.

My father's health was poor that fall and continued to get worse until he finally consulted Dr. McNulty and with his aid he gradually improved. February 6, 1898 he sent- for 19 volumes of books that he needed. Later in February he was working out one of his fossils.

>Tuesday, February 22, 1898
>
>As this was Washington's Birthday we did not have any school. I expected my books last night and I thought I would have a grand time today looking over and reading my books. But I was sadly disappointed, but managed to enjoy myself after all.
>
>Worked a good share of the day on the jaw and part of the skull I got on Sage Creek at "Crystal Hill.".. am getting it a little more complete all the time. I think my last summer's collection of vertebrate fossils is more valuable than I thought.

>Saturday, February 26, 1898
>
>Have received my books at last and am happy. They came last Wed. The lot weighed 150 pounds. They cost me $22.00 and the freight and expense was $7.35. Also received 9 pamphlets by Marsh and Cope by mail.

>Sunday, February 28, 1898
>
>Have been studying geology. Am gaining a little light all the time concerning my collection.

>March 6, 1898
>
>I have been enjoying life quite well of late since the doctor helped me out of my digestive troubles. I am spending nearly all my spare time studying my new books.
>
>I see that my business is not teaching in a high school, as I have not the interest in it one ought to have. For a week or two I felt blue about the school but interest seems to be reviving again and I must try to make it more intense. My heart is not in the work but I must get it through. My studies, I fear, are taking up a great amount of my interest. I wish I could have an equal interest in both.

Tuesday, March 29, 1898

I am beginning to rejoice with the thought of leaving. There is only one sad thought, it is leaving one person and perhaps never seeing her again. I thought I would tell her last winter what I felt about her but I suppose I will not and do not know what good it would do if I did.

I bought a harness of Mr. Foreman today, price $20.00. It is a good harness and nearly new. We now lack one horse, a tent, a wagon sheet, a saddle and a six-shooter. Mr. Coffey is going with me. I have so much to do this spring but fear I will not get it all done.

Sunday, April 24, 1898

I thought I would take a walk this morning as I am tired of being shut up in the house this beautiful day. I am now sitting on a block of basalt at the head of an old grave northwest of Virginia City. The sun shines warm and the weather is lovely. I can hear the hum of the flies as they fly from place to place. There are times when even the hum of a fly is pleasant. It is when we first hear it after the winter is gone. Ants are crawling around among the rocks, dry sticks and weeds as if searching for something.

The ridge on which I sit slopes away to the southwest toward Alder Gulch beyond which a ravine takes its place. A little way up the ravine, but farther to the left, is an old quartz mill. The ridge is composed of basaltic fragments. Some grass and some sage grow among the rocks.

There are several old graves here. One was once surrounded by a picket fence which is now broken down. These are probably the graves of some of the miners of the older days. Nearer to the city, just above town, are a number of graves and there the road agents are buried.

Saturday, April 30, 1898

I feel a little somber tonight on account of my financial condition. I hoped at least that I would have plenty of money to carry me through the summer but now it is a little doubtful. It seems I cannot save money. It does not seem as though I spend money foolishly yet it goes and I seem to have no power to stop it. I have earned $700, or will by next Friday, and I cannot start out with $100, and it is discouraging. Of course it costs considerable to start out; still the things I have bought are not very expensive. We, however will have a very comfortable outfit. I have a new saddle, saddle blanket, cot, tent, wagon cover, half of a good harness etc. We still lack a horse. Mr. Coffey will buy one I think.

My books cost a good deal, something like $60. But I know I will not have enough to last through the summer unless I sell some fossils or minerals.

Upon completion of the school term Father spent a few days packing and making ready for the summer's trip. He purchased an old mare from Mr. Cahill but before starting he became afraid she would not stand the trip so he traded her and $30.00 to Mr. Taylor for a bay horse. All his belongings were packed, no plans being made to return in the fall.

He decided to go to Bozeman before starting the trip and, with the aid of the recently purchased books, endeavor to determine the age and identity of the fossils from the Lower Madison. Before leaving a night was spent at the Goetschius ranch and he bade the family goodbye. On his way to Bozeman he visited his friends, the Huttons, near Ennis. Mr. Coffey, who was to accompany him on the trip, was teaching near Hutton's and they made plans to meet in Logan around the middle of June and start their trip from there. He visited the Hoelleins on the Lower Madison before proceeding to Bozeman, where he gathered up more fossils, the ones stored there, and took them with him.

At Bozeman Father's time was spent in intensive study of his fossils. Although no definite conclusions could be reached it appeared his discoveries were of an older age than he had supposed, and it seemed many of the species might be new to science. Among his finds were prehistoric horse, camel, rhinoceros, mastodon, oreodont, carnivorous animals and many turtles. This first real light on the fossils was, of course, fascinating and encouraging and he more than before realized the need of having fossils in a place where their study could be continued.

By June 25th he had met Mr. Coffey in Logan and they were ready to proceed. A vague plan was to go down the east side of the Missouri River below Three Forks for some distance then cross over the Big Belt Mountains to the east and examine the Smith River Valley around White Sulphur Springs. Princeton paleontologists had previously made interesting discoveries in this area. A further thought was to follow down the Smith River to Great Falls and then head westward toward either Missoula or Kalispell.

The trip was leisurely and quite pleasant. In the lake beds along the west edge of the Big Belt Mountains below Three Forks he found several valuable specimens of fossil mammals. They were having some trouble with the horses as is indicated in his diary.

> July 6, 1898
> Started to write at camp No. 9 when the horses started for home. We had to follow them about a mile and a half, they running with hobbles on. I made up my mind if they would not stop to eat, we would drive on until they would be glad to do so. When we returned we took down the tent, packed up and started on again. They had little to eat for about 10 hours and we thought surely they would be half starved and not offer to run for a good while.
>
> It was about dark when we started but the moon rose and we had a nice cool time to drive. Drove until about 12 o'clock, I think it was, then we stopped, put up the tent, and picketed and hobbled the horses.

They continued their journey northward, stopping in Tostin for supplies. When they reached Deep Creek they turned to the east following the creek where the main highway between Townsend and White Sulphur Springs now goes.

July 10, 1898

There is a line of ranches nearly to the canyon on Deep Creek. The lake beds end suddenly and the creek issues from a canyon of slate, and it is slate, slate, slate through the canyon across the divide and into the Smith River Valley.

The assent on the west side of the Big Belt Mts. is very long and in some places the road is very rocky. When the divide is reached the valley of the Smith River is seen and it is comparatively a short distance to the valley. The valley is high, about 5000 feet altitude. The land is fenced off, apparently without much regularity, into pastures. Very little grain is raised and the houses are usually far between. The ranches are owned by horse, sheep and cattle men who own large numbers of these animals. They are building a fence that passes by here that is 50 miles long and the man is said to own 30,000 sheep.

July 12, 1898 - Camp No. 14

Yesterday I walked about 10 hours searching the lake beds for fossils and finding nothing but a few worthless fragments of bone. From 9 a.m. to 6 p.m. I walked most of the time with nothing to eat.

This morning we started at about 7 o'clock, an early start for us. We came about 4 or 5 miles when Mr. Coffey found he had left a screw belonging to his camera so he got on Dick and went back and I got on Pete and went southwest looking over exposures in the valley. It rained but we hitched up and came on. I stopped and had a talk with Mr. Riggs who is collecting bones. (Mr. Elmer Riggs was with the Field Museum of Chicago, G.E.D.)

The trip continued down the Smith River. The camp of July 14th was near Fort Logan on the ranch of the man that established it. The fort had been abandoned about 18 years.

July 16, 1898 - Camp No. 17 on lower Smith River.

Night before last we camped on a little rushing stream probably 20 miles south of Milligan. Had a splendid place to camp. Caught a few fish. Milligan is a P.O. in a ranch and there are several ranches in the little valley in which it is situated.

A little before reaching Milligan the Cretaceous begins. . . I counted nearly 60 species of flowering plants besides sedges and grasses, a large variety. In some places along the road the grass was as high as the wagon wheels.

Monday, July 18, 1898 - Camp No. 19, Great Falls

Last night we camped on the bench at an old stock ranch. No one was living there. We found a stall for the horses and slept in the stable. The wind blew a gale. It blew hard all night and has blown hard all day.

When we camped on Smith River the wind blew the tent down and part of our bedding got wet.

Neither Mr. Coffey or I have schools yet. We have written several Co. Supts.

I do not like this country and do not care to stay here. Wish to leave immediately but want to get forwarded mail first. I think we shall go toward Missoula next or Kalispell. I have got out of the lake beds region and want to get back into it. . . Great Falls is quite a large place and is considerable scattered.

Friday, July 22, 1898 - Camp No. 21

We left Great Falls day before yesterday at about 6 p.m. At about 11 a.m. I went down to see the Co. Superintendent and got horses shod. Had to wait to see Miss Craven and then again to get my horses. Miss C. seems to be a very nice lady. She knew of no winter schools.

Sunday, July 24, 1898

We left camp west of Cascade Friday at about 3 p.m. Stopped to see about work. They said there was plenty of work in Cascade, none between there and Helena. Our funds are running low and we expect to work a while in haying.

Friday afternoon a man was killed by lightning on the same road we came. He must have been just ahead of us. His horse was killed also. He left a young wife and baby. They had been married about 2 years. I noticed sharp lightning ahead of us.

Wednesday, July 27, 1898

We began haying here last Monday. It is an excellent place to work. There is only once in a while a mosquito. The flies are not very bad yet. The weather is comfortable. The hay is excellent to pitch and not far to haul. The fellow we work for appears to be a fine fellow. We put up the hay at 60 cents per ton. Thought we would hardly make wages at it but guess we will, it is such excellent hay to handle. We have a nice place to camp. It is near Mr. Larmie's house.

August 10, 1898

We are now camped at Mr. Larmie's place below the Mission. Came down here last Friday to stack the hay here. This is not as pleasant as Sullivan Valley. I am laid up with a lame back. Yesterday forenoon it began to pain below the shoulder blades. In the afternoon I went out and on the second load I had to quit. Got so I could hardly lift an empty fork.

August 25, 1898 - Camp No. 28, northwest of Helena

Last Sunday Mr. Coffey and I hitched my team to Mr. Harris' buckboard and went to a picnic on the Dearborne River. There were quite a good many there. Had a big dinner all together. The men went fishing and caught quite a many fish.

While at the picnic Mr. Coffey received a letter from a member of the board in the district on Jack Creek near where he taught last winter telling him if he would go and make application he could probably get the school. He was anxious to go. I saw too that my hay pitching was over so told Joe

Larmie that we would have to quit. He was sorry to have us go and I was sorry to have to quit. We decided to break camp and pull for the South.

Thursday, September 1, 1898 - Camp No. 30
We arrived in Butte a little after noon Monday. Mr. Coffey left me there...was sorry to have him go. Butte is a very lively place. The mines there amount to something.

Sunday, September 4, 1898 - Camp No. 37, below Clinton.
Am getting along fairly well for the road. I am somewhat tired of nothing but narrow valleys and canyons and long for open country again. I hope to get to Missoula soon after noon tomorrow if I can awake so as to get an early start. I am anxious to get settled for the winter. I have feared a little lest I should not get a school. If I should not I would be in a bad fix I fear.

I am all alone but do not mind it so long as I am well. Sleep soundly nights regardless of wolves, mountain lions, highway robbers etc. My dreams are sweet and my health is good. My back is a little lame but does not bother to speak of.

September 23, 1898
Since writing last considerable that is interesting has occurred.

At Missoula I saw Mr. and Mrs. Libby and Bell, Mr. and Mrs. George Johnson and Mr. and Mrs. Moon. I did not stay long the first time but came up the Bitter Root and engaged a school. Went up the valley a days drive and then returned to Missoula. Visited there nearly a week. Had a pleasant time. I visited four departments of Central School Went to the University and became acquainted with Dr. Craig and Professors Smith and Elrod. Had a pleasant chat with Craig and Elrod.

I began school last Monday. Am boarding at Mr. Strange's. This is a lovely valley but I am not fully at home yet.

There is lots of fruit here. I do not think there is another so fine a valley in Montana.

Nothing in my father's diary gives any definite reason for locating in the Bitterroot Valley. There is no indication of any communication with the University prior to his arrival there. He had, before leaving on his summer's trip, written the agricultural college in Bozeman relative to storing and displaying his fossils there and perhaps working part time but little encouragement was received, their lack of money being the excuse. He had early in the summer written to the County Superintendent of Schools in Ravalli County, in which the Bitterroot Valley is located, and had received an answer listing a number of schools with teaching vacancies.

He apparently chose the area as being as good as any in which to teach, with the thought in mind that the University might eventually have something to offer. Professor Foster had mentioned this possibility soon after father had arrived in Montana and before the University had actually been established.

He now had two definite desires: first to find a suitable place for his fast-growing fossil collections, and second to find some kind of endeavor in his favorite studies, paleontology and geology, so that he would not have to teach a country school. As will be seen later, the move to the Bitterroot Valley was probably as advantageous as any he could have made.

⌘ ⌘ ⌘

A MASTERS DEGREE:
University of Montana

When Father met with the president and professors at the University of Montana, he found they were not only quite enthusiastic over his fossil collections, but were serious about making arrangements for storing and exhibiting his specimens. Although funds were scarce and the buildings not yet completed, they were interested in providing him a laboratory where he could continue his study of fossils. Since he was teaching in Victor in the fall of 1898, his hope was that by the fall of 1899 he would be working at the University. His diary at this time describes his activities.

> October 6, 1898 - Bitterroot Valley
> I have now been here nearly three weeks. My pupils are bright and learn readily. Boys are somewhat mischievous. Have a mile to go to school.
>
> Wednesday, November 2, 1898
> I have been communicating with Professors Elrod and Smith and am to have all my specimens go to Missoula. I am to give them part of my specimens and loan them part. They are to store the rest for me. So I will have all my specimens for Montana together. This will be a great relief to me. I have worried a great deal about them, but when they get there I feel they will be comparatively safe. These fellows at the University seem to be enthusiastic fellows. I expect to go to the University after my school is out and study my specimens and get them in good shape. I wish I could spend the winter there instead of here but wishing is useless.

Tuesday, November 8 1898 - Election day
This morning I went to Victor to vote. I am repairing bones. I had a pleasant surprise today. I was unwrapping the skull from the Deep River beds (Smith River), the one Mr. Coffey found. How I wished there was a tooth or two...Today when looking over the pieces, I found a tooth protruding from the clay and on removing it found four molars.

December 9, 1898
Returned this afternoon from the annual teachers institute. I see, as I have never seen for a long time, what a poor teacher I am getting to be. I must quit the business as soon as possible. I am getting encouragement in my geological work and I intend to stick with it and will evidently come out on the right side. The future looks brighter now at least. I think I shall teach only a couple of months more. I received a letter from professor Elrod today. The University wants me to do my work there. Elrod thinks they can print the results of my investigations and give me an M.S. Degree. I do not know what I will do.

Professor Elrod had gone to Bozeman and visited Mrs. Davis in Belgrade where some of Father's material still remained, packed and shipped the fossils to Missoula, at the expense of the University. Arrangements were also made later to ship the fossils stored in Virginia City and the Ruby Valley to Missoula. For the first time all of Father's collections were brought together.

December 18, 1898
I am longing to get away from here and get to the University but I must be patient. I think, too, a good deal about my next summer's trip and long to be out again.

January 1, 1899
Another year has begun. I might make a good many solemn comments and make resolves but they might not

amount to much. I am still alone in the world and "paddling my own canoe". I probably am not doing very well paddling; all the same, if a fellow paddles he will get somewhere sometime. Have not the interest in teaching that I used to have but am earning my living. I have not given satisfaction and thought of quitting and may do so yet. For the last few days I have been very uneasy.

Monday, January 9, 1899
I am beginning to make a collection of bones of modern animals. If I work up my material in Missoula, I will need a lot of such material. I expect to stay here not more than two months longer. Prof. Elrod writes that my room, the one they have saved for me, is a fine one and I am extremely anxious to get there.

January 22, 1899
Went to Victor and killed a magpie and owl for skeletons; today shot red-headed woodpecker and squirrel. Stewed skeletons to get bones. I'm getting wonderfully interested in my bones.

February 15, 1899
Have moved to Missoula. Was glad enough to leave the Bitterroot and to get to work here. I found them moving at the University but it was so I could go to work and have been unpacking since...I think I will give most of my material to the University, except the bones. I cannot afford to pack them around and I think they will buy quite a good many books for me.

March 3, 1899
It has taken a long time to get my specimens unpacked and arranged for study...I am studying camels now.

My father's health was giving him considerable trouble at this time. The doctor discovered he had kidney stones. Although his work at the University was progressing, he was handicapped by lack of money. This situation was relieved somewhat when Mr. Coffey offered him funds. Although he and Professor Elrod had expected to go to the Flathead Lake in May, the trip was postponed when he heard from his sister that his father was ill and growing weaker.

> Tuesday, June 13, 1899
> When I was at the depot in Helena waiting for the train, I saw Dr. Craig and Professor Hamilton. Dr. Craig said he did not know any man he would rather see just then; he told me he had something for me but could not give it to me until the next day. It was a diploma. ... We arrived in Missoula at about 2:00 a.m. In the morning I tried to get my luggage but couldn't, had no clean shirt. Mrs. W. found some old ones which had been done up but thrown away. She fixed one up and I wore it. I had to go down town to get shaved. Caught a ride in a delivery wagon, got shaved, went out in the street to look for a bus when I saw Lillian and Florence with the horse and buggy looking for me. Drove to the University, getting there at just about the appointed time. I was called into the room where the faculty, board and class of '99 were. The class had on their college robes and I felt rather out of place, but did not worry much about it. Soon we went to the chapel. A gentleman from the Anaconda Standard read the address and the diplomas were presented to us. I had the honor of being the first to receive the Masters Degree in the University. I have not been working for the degree, but for the love of the work, but the degree is very acceptable just the same.

> Monday, June 19, 1899
> I am at work on my thesis most of the time, getting it ready for publication. Professor Elrod has been typing it for me. We are making three copies.

The work was ready for publication June 22nd and handed to President Craig. Preparations for the summer's expedition were made June 23rd and departure from Missoula was early June 24th. Accompanying my father from the University were Professor Elrod, head of the Department of Biology, and Professor Smith, head of the Department of Geology and Mineralogy. Mrs. Smith was also in the party. They traveled by horseback up the Clark Fork River. They also had my father's team and wagon which carried their camp outfit. The object of the trip was to explore the area around Drummond for fossil remains. They arrived in the vicinity of Drummond June 30th and camped near a place called New Chicago, which is now non-existent. This proved to be rich fossil area and the party remained there until August 3rd. Another summer of fossil hunting had ended, one of his most productive in Montana.

> Sunday, July 2, 1899
> Yesterday afternoon we went to Drummond and made a partial examination of the Cretaceous. Mr. Taylor is making a tunnel, expecting to strike hard coal. His geological ideas, however, are new to me and evidently original. A little geology is a dangerous thing. I fear his hopes are doomed. Yesterday forenoon I was returning on Pete from a trip southeast of here and in passing through a muddy place, Pete fell on my leg pinning it between his body and a rail. I was lucky it did not break. Since I was very lame, I rode in a buggy with Mrs. Smith; Mr. Smith rode Dick. I found a large bone and fragments of others about three miles above Drummond. I do not hope to find fossils very abundant but may get enough to determine the age.

> Friday, July 7, 1899
> Thought we would get away before but we are camped here yet. Have not got the rhinoceros dug out and ready to ship yet and I keep finding things.

> July 14, 1899 - Camp No. 5 near New Chicago
> My best success has been since I wrote last. Day before yesterday, as my leg bothered me, I did not go out until 2

p.m. or after. I went to the hills near camp where fragments of bone were exposed and found three skulls and perhaps part of another. One broke into pieces getting it out but it can be put together. One place where there were a few fragments, I took out a large nodule with a pick and there was bone on it but I did not know what it was at first. The truth slowly dawned that it was the skull of a blastomeyrx (prehistoric deer). Perhaps a prize never made me happier.

July 15, 1899
Prof. Elrod still wishes us to go to Flathead Lake and we are expecting to go, though if I could I would rather continue my work through the summer as I believe I would reap a harvest and get nearly as much as I have in the past five years before.

Friday, July 28, 1899
The day before we took a load of specimens to ship and Prof. Smith went home to attend to business and to go to Flathead Lake where the Biological Station is located. Mr. McDonald came up the day before.

Apparently Father prevailed upon Professors Smith and Elrod to let him continue his fossil collecting for the remainder of the summer. It is not clear, but it seems that Mr. McDonald was a student at the University or had some connection there. At any rate, he accompanied Father for the rest of the summer. They soon finished their work at Drummond and were on their way to Whitehall in the lower Jefferson Valley.

August 11, 1899 - Camp No. 13, three or four miles west of Whitehall
We left our camp southwest of Deer Lodge August 6. In the morning we picked up the buffalo skull and mastodon tooth I found. The next night we camped near Warm Springs...We expected to get an early start in the morning. Mr. McDonald

went after the horses while I cooked breakfast. He did not come until breakfast was cold and then he came without the horses. I too traveled a long distance but returned without them. Later, we both tracked them a long distance east to a little valley where a man named Schmidt lived in a little cabin. We soon heard the bell. We swore vengeance when we had them in harness. We left camp at about 1:30 p.m. I sent the horses right through but we stopped some time after dark about five miles south of Butte to get some bread and potatoes.

Saturday, August 26, 1899

We are now in a camp a short distance east of Whitehall waiting for breakfast to cook. We have stocked up with provisions and are going northeastward. We got vegetables on the South Boulder and other things in Whitehall and have our wagon full. The South Boulder country is pretty fine, I think, though rocky in places. It has nice garden spots. I do not think much of the Jefferson Valley. The plains and benches are so thickly covered with cactus that it is difficult to walk. I received a check from Prof. Fred D. Smith and a money order from Dr. J. J. Buckley. Prof. Smith has now loaned me $35 and Dr. Buckley $15. I must now be in debt over $150 but I do not worry much as I think I will be able to pay it before long if I live and have health.

September 27, 1899

The indications are that my journal is dying a lingering death and is to be a thing of the past...There are so many things to which I shall wish to refer later. I drove to the Ruby Valley at the end of my trip...I had a splendid visit.

After visiting the Ruby Valley my father returned to Missoula and the University. Another summer of fossil hunting had ended, one of his most productive in Montana. His diary of 1899, especially the latter part, is sketchy and difficult to interpret, however a clipping from the Whitehall paper, The Zephyr, published August 1899, sheds some light on what was found in that area.

"Messrs. Earl Douglass and Homer McDonald, two representatives of the State University, who are making a collection of fossils, and making a study of the geology of Montana, were in Whitehall Tuesday. They are enthusiastic geologists and have found a fruitful field in this portion of Montana. . . . Primarily they find that this section was once a tropical zone and was infested with animals of queer formations and of monstrous size. . . . The two geologists struck the first field near New Chicago, from which they dug fossils and bones and shipped a wagon load of choice specimens. Near Pipe Stone, (Pipe Stone Hot Springs a short distance west of Whitehall, GED) where their recent exploration was carried on, they found in the large clay bluffs, . . . fossils of turtles that measured two feet across. . . . In this vicinity the gentlemen discovered fossils of ten or a dozen animals which lived in the age prior to that of the mastodon. . . . Mr. Douglass informed The Zephyr that the work was very interesting and that the search was being rewarded by a collection that will be deemed valuable to the University."

⌘ ⌘ ⌘

19

EXPEDITION OF 1900:
A Newspaper Helps

After returning to the University, my father made a short geological excursion. Upon returning from this trip he enthusiastically began unpacking and preparing the specimens collected during the summer.

> Friday October 12, 1899 - at Missoula
> My journal, I think will soon become extinct...What is the use of anything only to make others and myself as happy as I can while I live...My greatest ambition is to earn enough money to make my folks comfortable and do the work I love. Although I have not reached that time yet the prospect seems better than ever before.
> Last Friday, one week ago today, was one of the hardest in my experience. Prof. Bailey Willis of the U. S. Geological Survey, wished Prof. Smith and I to go with him to Bearmouth then across the range to the north and then down the Blackfoot and home on a geological excursion. He wished to settle some questions on geology. Each had a bicycle but I expected to overtake the team which carried the provisions and ride if I got tired. When we got near Clinton Prof. Willis went ahead to overtake the team but found that it was about 1 1/2 hours ahead so I was doomed to go with bicycle...I went on but for the last 10 miles I was terribly tired. Perhaps that was farther than I had ridden in all my life before. The road was up and down with rock. After that, however, we had a pleasant trip.

They were going to camp at Bear Gulch but they camped at Bearmouth so I had to return so I rode about 45 miles. I had several falls and some had an air of seriousness to them.

Friday, November 17, 1899
Last evening I attended a lecture on the genius of Shakespeare by Rabbi Eisenburg. It was very good indeed.
This afternoon there was a football game between the Agricultural College and the University teams. The Ag. Col. beat the Missoula boys 6 to 0. The Bozeman team has larger men but the Missoula boys did well.

Saturday, November 18, 1899
I find that this summer's collection is richer even than I supposed. The camel skeleton that we obtained southeast of New Chicago is near complete, though there are no jaws or teeth. The blastomeryx (prehistoric deer) skeleton near the same place is one of the best I have found so I can almost restore it. What I thought was mercochoerus (one of the oreodonts) is a new genus if not a new order. A fragment of a jaw I have been cleaning today proves to be not only a good part of a jaw but a good portion of a skull of perhaps a new genus.

December 9, 1899
The State Board of Education has met. Dr. Craig said they were well pleased with my thesis and they are going to print it. I have not heard from Prof. Willis with regard to printing my investigations as a bulletin. I wish it might be done.

With more and more fossils being discovered and exhibited at the University, interest in geology and paleontology started to grow rapidly in the state, this being an activity that had not been previously experienced in Montana. The arrangement appeared to be ideal. With my father's great enthusiasm and so much unexplored country for collecting fossils, and with a suitable place to work, the newly-founded university seemed destined to soon become an attraction to students of geology and paleontology.

A few quotations from a newspaper article, published in the Anaconda Standard of Butte, Montana, January 19, 1900 give an idea of the interest that was starting to develop.

> January 18, Missoula
>
> "One of the most interesting departments of the State University is that over which Earl Douglass presides, the Department of Historical Geology. Mr. Douglass is well known through his valuable work on the fossils of Montana and his contributions to geological literature pertaining to this section of the country, which is interesting and valuable. He is this winter engaged in the preparation of the fossils that he collected last summer, and as they are made ready for their places in the cabinet of the university their great value becomes apparent.
>
> Mr. Douglass is an enthusiast . . . and is looking forward to the time when the museum of the University of Montana shall be the important collection that, by virtue of its location , it ought to be. The start that has been made, chiefly through the efforts of Mr. Douglass, is a good one, and with a little co-operation on the part of the people of the state, the university museum may easily be made a splendid institution. The prehistoric record of Montana is said to be most complete. . . . In speaking of this subject Mr. Douglass said recently:
>
> "In Montana rocks are found that were formed in nearly all the different geological ages. These rocks contain fossils.
>
> "The extreme richness of some of the states has so absorbed the attention of scientists, that with the exception of perhaps three or four expeditions, Montana has been passed by.
>
> "For nearly six years I have spent all my spare time in searching the valleys of Western Montana for bones of mammals that once existed here. My collection now occupies a room in the university. It is by no means as imposing as

some collections, but it is full of scientific interest. There are something like 75 species represented by bones, skulls, jaws and teeth. Probably from one-third to one-half of them have never been described, and it will take years to determine all the species. ... My plan is to spend the summers in the field collecting fossils, ... and the winters in describing the new finds. The bulletins containing the descriptions will probably be issued, one, at least a year. One is ready for publication and will be in the hands of the printers soon.

"It ought to be a matter of interest and pride to our state institutions of learning to secure at least a share of these interesting and rare fossils before large expeditions come from the East and take them from under our noses.

"The Carnegie Museum, which has now entered the field of exploration and the American Museum of Natural History with the enthusiastic and scholarly Henry F. Osborne at its head, is pushing the work energetically.

"We have not the money that they have, but we have the advantage of position, and expeditions can be sent at a fraction of the cost. For 20 years Yale College received the greater part of the fossil wealth of Wyoming. Five years ago the struggling University of Wyoming determined that a state museum should be established, founded upon the fossil products within the state. ... Today this young college stands second only to Yale among museums of the world in wealth of reptilian remains. It is to be hoped that the young and struggling University of Montana will take similar steps soon, and I believe it will. The university can be a help to those who have an interest in the rocks, fossils and minerals of the state, and those people can, in turn, help the progress of the university."

The future appeared bright at this point but the important work of preparing the fossils was now being hampered by my father's financial circumstances. He had been operating on borrowed money for almost a year with the hope that something would develop at the university enabling him to carry on his work and earn a living at the same time. The University did the best they could by giving him a part-time teaching job but this was of only a temporary nature and did not provide any definite plan for the future and took much of the time he had hoped to spend on the fossils. Things began to look a little bleak for the future, especially for the next summer's fossil collecting expedition. As the diaries reveal, two men, Mr. Stone, of the Anaconda Standard, and Professor Scott, of Princeton University, were soon to enter the picture and influence the course of events.

December 10, 1899

I am boarding at Mr. J. H. Kennedy's. Have not so far to walk as I had last spring. I enjoy the work far better than that of teaching public schools. . .I like the studies I teach except physics and that I like except the part we are on now, mechanics.

Friday, December 22, 1899

Today is an important day in my career. I have eaten the flesh of the far famed but almost extinct American Bison. I had often wished that I might have that privilege.

Monday, February 5, 1900

Have been at work at the University today. Mr. Ebert and I are arranging, numbering, and cataloging rocks.

Prof. W. B. Scott writes that he wants to use my new material in a comprehensive work on the mammalia which he and Prof. Osborne are working on so he hopes I will publish soon.

Saturday, February 10, 1900

Have been at the University today boiling and cleaning bones. I received some time ago a good share of the skeleton of an antelope from Billings. The back part of the skull was shot to pieces and the feet were cut off yet I was glad to get it.

Saturday, February 24, 1900

Have been up to the University studying my fossils and preparing bones. Described a new species...I have now described only 4 species and I have only a little over 3 months left but will hope I have more time in the future.

Sunday, March 4, 1900

I was up to the University yesterday also. There is so much to do besides this work. I have to teach geology, physics and physiology with at least three afternoons occupied with laboratory practice. I hear that one of the professors wants me to have the other days occupied. If the president insists that I must take the laboratory in physiology I will not have much time except Saturdays and Sunday, I fear. I do not think I should be expected to do as much as the other professors who get 3 and 4 times as much.

Saturday, March 17, 1900

Professor Smith told me Wednesday that Mr. Stone wished to devote a page of the Sunday edition of the Anaconda Standard to my work and have it illustrated. Yesterday Professor Elrod and I were engaged in photographing my room and some of the more interesting fossils.

Mr. Stone came up to see me yesterday morning and I had a short talk with him. He wants to boom the thing and get money to pursue the work.

I do not think the University can help me, and do not think they feel that they can spare the money. If my expenses are paid I expect to collect for the University, that is if I return another year, which I hope I may be able to do.

Friday, May 25, 1900

I have felt very blue about my future as it was so uncertain. At last I was persuaded to go and have a talk with the president. After I had done so I felt much better. I think now that I will return. He says the only question is the financial one. He commended my work not only in the line of investigation but in teaching. I hope they will manage in some way to give me a job and enough to live on. I think there may be an opening in a year or two.

Monday, May 28, 1900
I expect now to go collecting this summer for the University. Mr. Stone of the Anaconda Standard and Mr. Dirkson raised $250.00 for the purpose.

Friday, June 8, 1900 - Potomac P. O.
Beginning of geological expedition. Started from Missoula this morning on my summer's trip. Bought a horse at Bonner a few days ago, and am following out my plan of going over to Laurin by way of the Blackfoot with one horse. Got an old cart of Mr. J. H. Kennedy.

The State Board of Education voted to retain me for another year. I had a talk with Professor Hamilton and he said that they put it down at the same price as last year but they would pay in proportion to the work. He had heard about my being offered a fellowship at Princeton and advised me to take it as it was better than they could do by me or at least he intimated it. I expect now to go and if I do not I shall be rather disappointed.

The Blackfoot River enters the Clark Fork River from the north, a short distance east of Missoula. My father went up the Blackfoot River instead of up the Clark Fork River as he had done the previous season. He had left his team and wagon in the Ruby Valley when he completed his trip the fall before. Apparently Pete had died during the winter, this being the reason for his buying another horse. As he proceeded up the river he passed through the little towns of Potomac, Ovando, Helmsville and Avon. Not far from Avon he lost his horse.

Saturday, June 16, 1900
Monday night I slept out of doors about 7 or 8 miles north of Avon. Got up in the morning and went to catch my horse

and he suddenly started on a run with the hobbles on. I did not think he would run far with the hobbles on and started on the run myself. But he followed a band of horses and went to the top of a hill. When I got there, they were out of sight.

The search for the horse continued all day. He said it was not entirely in vain for he found some fossil shells he would not have otherwise found. He eventually borrowed a horse from a Mr. Rice, and after considerable more searching, the lost horse was recovered, his feet being so sore from the hobbles that he could scarcely walk.

The bones of an oreodont and other fossils were found in the vicinity of Avon. Upon leaving Avon, he proceeded south crossing over to the Clark Fork River drainage in the vicinity of Deer Lodge. From Deer Lodge he continued on south, across the continental divide to the town of Melrose. He noted that the lake beds continued across the divide. He found more fossils in this area and at Melrose shipped 100 pounds to the University. From Melrose he went east to the Ruby Valley, arriving there the 21st of June. Here he met a Mr. Murray, whose identity is not clear, but who undoubtedly had some connection with the university. Mr. Murray was to accompany him for the remainder of the summer.

> Friday, June 22, 1900
> At Mrs. Van Broklin's in the Ruby Valley.
> I found Dick as fat as a pig. . . . In the evening I went up to Mr. Taylor's. They urged me to stay all night. I had a long talk with May. She heard that Pearl and I were to be married.

> Sunday, June 24, 1900
> Thursday I spent most of the day mending my tent and wagon sheet and doing little things. Called on Mr. Goetschius' people in the forenoon and again in the evening.

After Mr. Murray arrived they moved to the Upper Ruby Valley and examined the lake beds there for a few days, after which they crossed over to the Madison Valley by way of Virginia City and camped on

Wigwam Creek. A rancher in this area, by the name of Thompson, had found bones near Black Butte, the highest peak in the Gravelly Range to the south. The Gravelly Range forms the divide between the Madison and Ruby Rivers. He thought they were the bones of a mastodon, but Father had examined the bones on his first trip to the Ruby Valley in 1896 and was certain they were dinosaur bones. It would be a rugged trip into this wild region but he now decided to attempt it. In those days the only practical route to Black Butte with team and wagon was to climb to the top of the range near the head of Wigwam Creek then travel the top of the range in a southerly direction some 20 miles. Today this is a delightful drive over a good graveled road through a very scenic country but in 1900 there were no roads. The highlights of the trip are recorded in his diary. In a previous entry he had partially told of their predicament while prospecting on the Upper Ruby Valley. Dick, his horse previously mentioned, had somehow gotten into a mud hole and it was with difficulty and anxiety they had finally gotten him out.

> Sunday, July 8, 1900
>
> In camp on Wigwam Creek above Harry Thompson's ranch. It is horribly hot and I am suffering from sunburn so it is not very pleasant. There has been a scarcity of rain all places I have been since I left the Blackfoot.
>
> This I believe is the first day that I have not been hard at work since I started. ... We went down to Mr. Thompson's and took dinner.
>
> In my last entry (when camped on the Upper Ruby) I started to tell about Dick's misfortune. We had a hard time getting him out of the mud and I feared we never would get him out. I thought two or three times he was near his end. We hitched the cayuse to him but after pulling two or three times he would not pull any more.
>
> After this we got into a place where probably a wagon had never been before. We went along a ravine where the water had cut steep sides, and came to a place where we could not go ahead, and we could not turn around

either, so we backed the wagon out by hand, unloaded, unhitched, turned the wagon around, loaded up and went back taking another gulch.

Monday, July 9 1900

Am now sitting by a camp fire on a beautiful piece of grassy ground by a wooded depression or ravine nearly at the top of the Gravelly Range. The fire is snapping and sending up sparks into the air which is now becoming dim with night. In the east, dim in the approaching dark, is the high rugged Madison Range. The mosquitoes are singing around my ears. Now ashes of sparks that have ascended are descending on my book. To the south east the moon stands above the grassy slope. If one looks in that direction or to the south it looks like the prairies of Dakota. . . . To the southwest are a couple of horses feeding and a bell is jingling. They are the animals that brought us up here to the top of the world.

We started from Mr. H. Thompson's about 1:30 . . . came a round about way to get here where there is no regular road.

Thursday, July 12, 1900

We are now camped by a sheep camp on the Gravelly Mountains several miles from where we were when I wrote last. We have been to Black Butte and are now returning. Tuesday we passed the last sheep camp and this year's wagon tracks. We stopped to inquire the way of Mr. Kennedy a herder. We found the way from here bad in places. There is a long steep hill to pull up and we then went down a very steep incline covered with rocks. Soon after this the reach (a pole connecting the two axles of a wagon, DDI) broke in two. We were far from anyone, or where we could get any thing fixed. It was not a very pleasant situation. Fortunately we were near wood and water. . . . I got supper while Mr. Murray went for water. After supper we got the box off. I took a piece of pine board and using a knife and hot tent pin for an auger, got some holes through it, put some rope around, putting a twist in it, and it held to go over some terrible rough roads, or no roads at all.

The next morning I took Dick and rode over to where Kellog and Thompson found the dinosaur to look for a suitable camping place. The flies and mosquitoes are very bad now. I never have seen the large horse flies so numerous before. I wanted to find not only wood, water and feed but a thick woods where the horses could escape the flies. ... I found a good place I thought. Did not succeed in finding the dinosaur or any fossils of any kind. Was considerably disappointed in the place. Before I got back to camp I concluded we had better not try to drive there with the wagon not knowing whether the reach would hold or not. I lay the matter before Mr. M. He was about played out, but seemed to be willing to go, but I did not want to insist. However, when we started back he made the remark that he would liked to have gone, so we turned around and went back.

We went over and camped near Black Butte but stayed there only one night. The flies and mosquitoes were very bad. The next day we returned to the sheep camp which Mr. Tom Kennedy attended.

About three days were spent in the vicinity of the sheep camp after which they returned to the Madison Valley. Father expected to find money, from the fund raised by Mr. Stone, waiting for him as soon as he got his mail. This was not the case, however.

> July 22, 1900 - Madison Valley, Montana
> I expected to collect another day on Wigwam Creek. I also expected to get a check for my last two month's salary. Went over to Mr. Miles to get my mail but no check. I confess I was pretty wrathy. It is an abominable shame. I concluded I would send to Mr. Stone for it. We packed up and went to Ennis where I telegraphed Mr. Stone.

July 25, 1900
As before, I was disappointed in not getting any money. I sent word to Mr. Stone to send me money quickly but it has not come.

Mr. Chowning, had the evening before, offered to loan me anywhere between $25 and $100. I concluded I could take $15 and it is a good thing I did.

Mr. Chowning was a merchant in Ennis, an old acquaintance. This was the beginning of an awkward situation between, my father, Mr. Stone and the University. Apparently, Mr. Stone, upon hearing that Father might go to Princeton, became suspicious about receiving the fossils being collected if the money was delivered too soon. It probably never entered my father's mind that Mr. Stone would doubt his honesty. The professors at the University tried to help but apparently could do nothing.

With the $15 borrowed from Mr. Chowning, Father and Mr. Murray left Ennis with the idea in mind that they would travel eastward and on their way examine the Musselshell river country north of Big Timber and east of the Crazy Mountains. Mr. Murray's destination was Miles City so this plan was agreeable to him.

They first visited the Lower Madison Valley, Logan and Bozeman, where considerable delay was caused by not receiving the money they needed to continue the trip. The time was partly utilized by collecting fossils near Logan and Three Forks. Part of the money was finally received July 28th and they continued their journey, passing through Bozeman the 29th and Livingston the 30th. August 1st they were in Big Timber.

Wednesday, August 1, 1900
We had given up the trip to the Musselshell. . . . On account of being delayed in getting the money and making side trips I had concluded to drive on toward home probably as far as the Black Hills, if I did not find good collecting along the way. I thought I might sell my outfit perhaps for enough to get to Princeton.

While we were stopping in Big Timber, someone showed Mr. Murray a piece of a dinosaur bone from the Fish Creek country. Mr. Murray called my attention to it and I went into the Meat Market where it was and got to talking with the men there. They thought it would be a good country for collecting specimens. I wanted, if possible, to get the $150 still due me, and wanted at the same time to get a collection more than worth the money. I thought if I could get a fine collection, including dinosaurs, they would be almost obliged to send me the money.

Fish Creek was north of Big Timber, not far from the Musselshell River where they had originally planned their trip. They arrived there August 3rd and the next day a Mr. Albert Silberling showed them some deposits and they dug out about 25 vertebrae of a dinosaur. They left the bones with the intention of picking them up later and spent a couple of days prospecting. When they returned to prepare the bones for shipping, they found to their disappointment, they were all gone. They found a freighter had picked them up and he would not release them until they had proof of who had dug them out. After much trouble they finally got them back. After this incident a good collecting area was located about seven miles north of McClatchey's ranch and the greater part of a dinosaur was taken out.

October 3, 1900 - Princeton, N. J.

I hope, hereafter, I shall have more time to write in my journal. I guess before I review the past I will have to get a calendar. I think we camped on the Mcclatchey Ranch about 7 weeks.

Some of the later additions to my collection were some nice fossil leaves and portions of a mosasaurus. ... The latter has the skull and the jaw and a lot of the vertebra. It is a fine specimen.

About the last of August Mr. Stone sent me a check for $50. I thought I would return it, but it seemed the only way to get home so I took a load of specimens to Harlowton and then got ready to leave, pulled down my tent and went to Big Timber with Mr. McClatchey, and took the night train for home. I was at home 9 days; left there Thursday, Oct. 4. Anyway I arrived here last Tuesday

I went in the evening and saw Professor Scott. Had an interesting talk with him. He seems a quite serious man but very kind and exceptionally fair minded. Everyone seems to like him. He is an exceptional man.

⌘ ⌘ ⌘

20

PRINCETON UNIVERSITY:
A Storehouse of Scientific Knowledge

The unpleasant hassle with Mr. Stone over the finances for the summer's collecting trip was to be somewhat compensated at Princeton University. Here my father was to study under the renowned paleontologist, William B. Scott, and would be able to build a sound foundation for his future in paleontology and geology. Also, through his association with Princeton, he was to come in contact with many of the prominent men of the time in these fields. However, the struggle for enough money on which to exist was to continue.

> Arrived here Oct. 9 in the afternoon about 4 p.m. . . . Went to Nassau Hotel. The next day went to the office to see about a room, and the treasurer gave me a check for $100. This removed a burden from my mind as I had only $10.00 to start with and did not expect to get money at once. Professor Scott, the night before, stopped at the hotel to tell me I could get a room in Edwards Hall. Professor Farr came with me and I chose room 27 South Edwards. Began to try to get things to furnish it. Bought bed, table etc., and slept in my room that night. Have my room now so it is quite comfortable except the stove does not work good and am afraid I will have to discard it.
>
> Besides my other work, am taking Professor Scott's Lecture in Paleontology and Historical Geology, Professor Maclosky's Lectures in Osteology and McClure's work in Vertebrate Anatomy.

October 26, 1900

They have a fine collection of modern skeletons, one of the best if not the best in America, so I am trying to make the good use of my opportunities in this line. Am boarding at Mrs. Leigh's on Mercer Street.

This is certainly an interesting change for me. Almost everything, in a sense, is new. So far I like the college better than any other I have attended except the University of South Dakota at Vermillion. That was only a fraction of the size, but that was a short golden experience of my college experience, an age that will not return.

This is a pretty town one of the prettiest I have seen. The college grounds are nice. They have many nice halls but not enough room for the educational part.

I have occasional talks with Professor Scott and it seems there is much interest in my collection. I am anxious for it to come, at least part of it, so I can go to work.

October 30, 1900 - Princeton, N.J.

Day before yesterday was my birthday, but I forgot all about it until next day. Only to think 38 years and still living. It seems almost like hundreds of years. And what a strange life it has been, it seems to me. Here I am in the neighborhood of 40 and still struggling to get an education. The riches that every boy thinks will be his for this age have not yet appeared. About all I possess are books, bones, a team and wagon etc. I am about where I ought to have been 15 years ago. But I am a student because that is all that satisfies me.

Sunday, November 4, 1900

Yesterday was the Cornell-Princeton football game on Princeton grounds. Cornell won the game by a score of 12 to 0. Princeton feels blue and it is extremely quiet around here today. . . . They feel afraid of Yale now.

Wednesday, November 7, 1900

Yesterday was general election and it was quite lively in town last night. I went to bed but did not go to sleep and heard a lot of racket over on Nassau St. so I got up, dressed and went over. There came near being several fights. Mr. McKinley is elected by an overwhelming majority. They had a negro band and a procession, mostly negroes, marching through the streets.

I received my first box of specimens yesterday. Professor Scott and Professor Farr think they are very interesting. When Professor S. saw the pig from the Flint Creek beds (near Drummond) he said, "That is the first American pig. It is a true pig."

I have not received the last $100 that I was to have for my last summer's expedition. Scott sent me $50.00 to Medford to get here. I spoke to him about it today and he said there was no hurry at all, to take my time. He is an exceptional man and I seem to be in luck.

Friday, November 16, 1900

Am getting my fossil rodents in shape for description and am studying them some. Have something like 15 species I think. Several of them are evidently new.

Met Professor Osbourne today. (Henry Fairfield Osborn, Curator of Vertebrate Paleontology, American Museum of Natural History. Famous for his book "Men of the Old Stone Age" SAB). I took a particular liking to him. He was at the meeting of the Scientific Club. Professor Scott gave an exceedingly interesting lecture on the Mammals of the Santa Cruz Beds.

Some fellows thought I was a Freshman and came pounding at my door this evening. I thought one or two were drunk and would be unreasonable so did not want to let them in. They began to batter down the door and I opened it. One saw me and knew who I was and did all he could to make things right. One was evidently considerably under the influence of intoxicants.

Sunday, November 19, 1900

Yesterday was the great event of the season in Princeton – the football game between Yale and Princeton. There were I believe 20, 000 people in attendance. Many special trains bore the thousands from New York, Philadelphia and other places. My purse had run low and I was in one of my fits of despondency and fear of the future. The money I expected has not come and it leaves me in bad position again. But I ought not yield to discouragement and whining.

But I was going to say I wanted to go to the football game. When I saw the throng of people and realized the momentous occasion I could not rest. I had little money in the bank, only enough to pay my board for a week or two. The admission was $2.00. I, however, changed my clothes, went out to Nassau St. where they were selling tickets for $2.50 and $3.00 each. I went down to the grounds, went around to the S. side and looked through the fence for a while. I was not used to that sneaking way but could not resist. Finally after the game was in full tilt I got a ticket for $1.25 and went in. I did not take an extra coat so sat and shivered but had the pleasure?? of seeing Princeton go down 29 to 9. At least it lent variety to the day. I expected to catch a bad cold but did not.

December 24, 1900 – Princeton, N.J.

I have all but given up keeping a journal. . . .

I have been happy most of the time since I last wrote. My financial embarrassment has continued but it has not caused me great anxiety but once or twice. Once I was blue for a little while when I ran short and had no money to pay my board. But when I went to Mrs. Leigh, and told her the circumstances, she said it was all right, to not worry about it and pay when I got it. When I found she was not going to run the club through vacation, Professor Farr offered to let me board there. He has been extremely kind to me since the start. I hope my deeds show my gratitude.

I received a few days ago a letter from Professor Hamilton (of the University of Montana) which to me was surprisingly encouraging. They expect to establish a department of geology and Professor H. says he is personally in favor of my having the position. I do not think anyone on the board would oppose me unless they had some personal friend that wanted the position. I thought Professor H. was in favor of me last spring but I feared that trouble with Stone would be dangerous to my interests.

December 25, 1900

Took dinner with Professor Scott and family today. It has been a pleasant Christmas.

Mr. Hatcher, from the Carnegie Museum, (J. Bell Hatcher, SAB) was here quite a while 2 or 3 weeks ago. I talked with him about the fossils I collected near the Musselshell last summer. He said if I concluded to sell them, he would pay the freight to Carnegie Museum.

December 26, 1900

Have been getting fossils ready for study today. At work at rodent skeleton. Have been studying the tapir and rhinoceros like animals, but did not get any named. Got exceedingly interested. Am anxious for evening to come, so I can continue working out the puzzle.

January 1, 1901

Well the new century is getting along fairly well so far. Seems a good deal like his dad. In fact if I had not kept track of the time I would not have known that the old fellow had stepped out and the young one stepped in.

I am tired today on account of sitting up late last night. Got up too late for breakfast.

Have been at work as usual nearly all day. Professor Farr and Professor Scott have been working too. Lately I have been getting material ready for study. Have not written a line for perhaps a week or more. It seems that every thing I take up and study is different from anything that anybody else has found.

I wish I could go to Montana and continue the study of the lake beds next summer. Perhaps I can.

January 8, 1901

Yesterday I received a letter from Pearl. (Pearl Goetschius) I was glad, yet sad. She has been spending her vacation at home and returned to work at the Orphans Home in Twin Bridges. (Twin Bridges, Montana, GED). The Dr. told her she had been exposed to small pox. It will be an awful shock to her mother. I may not hear from her for weeks.

Queen Victoria died this afternoon. She was 81 last December I think and had reigned almost 64 years. It was one of the great reigns of history.

Sunday, January 27, 1901

Pearl is all right. She did not have the small pox, and I suppose is back teaching now. I am so glad. I worried about her.

In March Father took an excursion to Washington D.C. where he looked up an old Brooking College acquaintance, a Mr. Wilcox, who was then connected with the Department of Agriculture, and together they took in some of the sights. While there he contacted a number of men connected with the U.S. Geological Survey and the National Museum. These were prominent men of the time in the fields of geology and paleontology. He also got in touch with Mr. Weber, his botanist friend, with whom he had worked while in St. Louis.

April 7, 1901

Again after two months and more of silence I make another effort to write in this book. It seems that when I need to keep a journal the most, I never get time. It has been quite an important time for me.

The night of the second of March I went to Washington. Got an excursion ticket for something over $5.00. Got there Sunday morning. After being taken to the wrong place on the street car, and walking a good deal of time, I got out to Tacoma Park and found Mr. Wilcox. The next day we attended the inauguration. We pretty nearly got mashed.

Monday I went down with Mr. Webber and met several botanists. In the afternoon I met Stanton, Knowlton, Ward, White, Lucas, Peale and others.

Later in April he visited New York City where he met Dr. W. D. Mathews, (William D. Matthew, SAB) Associate Curator of Paleontology, at the American Museum of Natural History. Had it not been for illness this visit to New York City should have been enjoyable and instructive. Of this trip and other matters important to him he wrote in his diary.

April 14, 1901

A week ago yesterday morning I went to New York City. Looked up what I wished at the museum but was sick and took the 4:55 train for Princeton. Met Mathews, (Matthew) Hay, Gidley and others. Mathews (Matthew) asked me to take supper with him and stay there over night and I wanted to go to the Buffalo Bill's Wild West Show and visit the zoological gardens, aquarium etc.

Went to Trenton week before last and got a suit of clothes and overcoat. Have been in such straightened circumstances all winter that I was not able to get an overcoat but wore my Macintosh all winter. It was warm enough, but no one else seems to do it here though some do not seem to have overcoats.

The matter is settled between Pearl and I. It only remains when? It makes me happy. There is only one thing that bothers me much. . . . I do want to be worthy of her and do not want her disappointed.

I have given her another chance to escape and she does not accept. By what she writes it would be about the hardest thing she could do.

May 11, 1901

There has been a great change in the aspect of nature since I wrote last. The woods are green, the greater part of the trees are leaved out. Many flowers are in bloom.

The last two Sundays I have been out walking. Guess I got my face poisoned a week ago. It has been breaking out and causing some trouble.

I have finished my paper, "The Mammalia of the Montana White River Beds". Mr. W. Nelson made figures for one plate illustrating the 4 genera and one species. . . .

Professor Scott told me a day or two ago that I could have the same fellowship next year if I wished. They never have given it twice but the fellow who was appointed resigned. It seems lucky for me.

Professor Osbourne (Osborn) was here yesterday and I got better acquainted with him.

Heard Mr. S. L. Clemens (Mark Twain) lecture night before last and last night at Princeton-Harvard debate. He is funny surely and I am happy that I was lucky enough to hear him.

My Pearl is faithful and kind and true and affectionate. It is settled between us that we are to share each others fate but she wants to go to school another year, and I want to get a good position. I probably shall return to Princeton but to think of being separated from her, more than another year, makes my heart faint, but it must be. I want to get a position in Montana.

⌘ ⌘ ⌘

PRINCETON EXPEDITION 1901:
A Remarkable Discovery

After receiving a letter from Professor Hamilton of the University of Montana regarding the establishment of a geological department, Father apparently entertained hopes of returning and wrote Dr. Craig regarding the matter. The answer to his communication, dated April 5, 1901, reads in part as follows:

> "Both of your letters have been received. In regard to the position in Geology, I am sorry to say that has already been filled by the appointment of Professor Rowe of the University of Nebraska. He was elected in December, as instructor in Geology and Physics.
>
> If possible it would be very pleasant to me to make some arrangement that would attach you to the University. I will consider the matter, see what can be done, and write you later.
>
> Yours very truly,
> Oscar J. Craig, President"

When it was found that another fellowship could be obtained at Princeton apparently nothing more was done relative to his becoming further connected with the University of Montana. From an article published in the Billings Gazette, January 30, 1966, it seems the field of paleontology experienced a long period of inactivity after 1901. A part of the article follows.

> "Western Montana holds treasure more precious than gold for the vertebrate paleontologist. And now, after more than a decade of rugged spade work, Dr. Robert Fields of the

University of Montana has an honest-to-goodness, first rate curriculum offering in the fossil sciences.

It's ironic," Dr. Fields said the other day. "When I came here in 1955 the university had just three vertebrate fossils. Yet the first master's degree awarded by UM in 1901 was in the field of vertebrate paleontology." This first graduate degree, listed as Thesis No. 1 in bulletin No. 1 of the graduate school, went to Earl Douglass, who later went to work at Carnegie Institute.

Who knows what caused the 54-year lag in a field so basic to understanding the earth's evolution? Montana's traditional preoccupation with practical, "cash money" affairs likely checked any serious efforts to probe for fossils.

But Dr. Fields' arrival signaled a new era. Paleontology was accepted and fully approved."

Sometime during the winter or spring of 1901 plans were made, probably by Professor Scott and my father, for what was known as the Princeton Expedition of 1901, its purpose being to further explore the region of the Musselshell River where father had experienced fair success the previous fall. He was anxious to return to Montana in the spring of 1901 for numerous reasons. He wished to continue the fossil collecting in the area of the Musselshell and part of his last year's collection was still stored at McClatchey's ranch on Fish Creek awaiting his return. He had not yet received the $100.00 owed for his previous summer's work, which Mr. Stone was holding. But perhaps of most importance was a trip to the Ruby Valley to visit his sweetheart, Pearl Goetschius (my mother). In May 1901 he was anxiously awaiting the start of the expedition.

> May 15, 1901 – Princeton, N. J.
> I am now waiting with some impatience to go home. Have been delayed on account of R.R. passes. Can get the round trip for about one fare from St. Paul to Big Timber on the Northern Pacific.

May 28, 1901
I started from Princeton yesterday afternoon at 4:30 p.m. I had expected to start from here between 2 and 3 weeks sooner so the delay was unpleasant.

June 25, 1901 – Big Timber, Montana
Since writing last, nearly a month ago, I have had considerable experience and time to write, but have not had the disposition.

I stayed at home until Monday June 17 when I met Professor Farr in St. Paul, and that night we departed for Montana. Arrived here the morning of the 19th.

Thursday the 20th, in the evening, I started for Missoula. Lodged for the remainder of the night in Missoula at the Chapman Hotel. The next day I saw Dr. Craig on business. Also met Mr. and Mrs. Kennedy, Nettie White, Mrs. Craig and others.

The matter of the fossils was settled by Mr. Stone paying me the $100 due for collecting last year. It has caused me a great amount of trouble. I was glad to get the money. I had wanted to stop and see Pearl when I first arrived in Montana, but did not have as much money as I wanted and could not get a buggy at the livery stable, so when I got the money I decided to stop and fixed out in Whitehall.

Whitehall was the nearest train stop to Twin Bridges where my mother was then teaching. After renting the team and buggy in Whitehall, Father drove to Twin Bridges, a distance of some twenty miles, where he joined her. Together they journeyed to the Goetschius ranch, another twenty miles. He mentioned the trip being very pleasant, especially when they were alone in the buggy. After a short stay at the ranch he returned to Big Timber where preparations were being made for the summer's fossil collecting expedition.

June 26, 1901

We bought six horses, four only recently broken. I fear a little trouble, for the boys should be fairly good riders for these horses, but it was the best we could do.

July 6 1901 - Camp near Fish Creek

We left Big Timber June 27. The first night we camped on Wheeler Creek south of Melville. Some of the boys had trouble with their horses. Mr. Dugro got thrown but no serious damage was done.

The next day we camped by a spring on the road to Middecombie's. We stayed there Saturday and Sunday.

Monday we moved down to this camp. It is south of McClatchey's house on a little creek by a spring. It is a pretty good camping place.

The country is beautiful now as everything except the exposed rocks is covered with verdure. There are many flowers in bloom. I wish I knew the names of them, but I know but few.

The Princeton Scientific Expedition of 1901 was now underway. The party consisted of Professor Farr of Princeton, a group of young Princeton paleontologists, and my father. I have heard Father relate some of the experiences of the trip, especially when they started out with the wild half-broken horses, which were all that were available. Most of the boys from Princeton had scarcely seen a horse before. According to him the first few days were as good or better than watching a three-ring circus. Fortunately no one was seriously injured.

They found a few good turtles and numerous dinosaur bones, most of which were in a poor state of preservation and could not be saved. During the latter part of July Father and Mr. Willis, one of the young men just mentioned, started on a rather extended trip, the object being to trace some of the formations further east to determine their extent and to see if some better material could be found. The rest of the party stayed in the vicinity of Fish Creek.

Saturday, July 27, 1901

A week ago last Thursday, July 18, Mr. Willis and I started on a prospecting trip, since not much that was good had been found here. My plan was to follow the Pierre shales and the gray dinosaur beds that underlie them to the eastward. That night we stopped at Mr. Flanigan's. . . . They treated us royally. We had a nice bed and a fine time.

Friday I started on, while Mr. Willis went up on the bench to see some lakes. He saw wolves, antelope, ducks. . . . I found part of a skull. . . . Went down to Mr. Bromley's ranch. Mr. B. is a state representative. We ate supper there then went on and camped in a small deep ravine after dark.

Saturday we started from Forsyth's. At Forsyth's inquired the way to Whitney's. . . . Followed a ravine and found a good bone and some fragments. Mr. Willis shot at a band of antelope several times but did not get any. We went to Bustein's and got supper. The men there said there were big bones up on the bluffs. . . . We found lots of fragments but no good bones. We stayed there over night.

Sunday we went over to the next house S.E. We concluded not to go any farther, but return and look over the stratum thoroughly that Mr. Bustein told me about.

Monday we started to return, but as we had no food Mr. Willis thought we ought to go across the basin to Mr. Sharp's so we did so. . . . Mr. Sharp's folks were not home but we got some biscuits and milk which lessened our hunger. . . . We returned to the place where we found the good bone a day or two before. In a little ravine that I had not been up before, I found part of a skull and skeleton of a dinosaur. Only dug out enough to be assured it was good. We had been somewhat discouraged or disappointed, but that made us feel happy. . . . We camped after dark by a cabin and sheep corral.

Tuesday we looked over some exposures, but found nothing of importance. We took dinner at Mr. Forsyth's. He was not home. . . . Miss F. directed us the wrong way and we went across hills and through fences and finally got to Tudor's, then Flanagan's. Stayed there all night.

Wednesday we cut across the hills for Ferrington's. . . . Dick lay down and I feared he was going to give out. It was getting dark and we were wondering where we were going to spend the night when we heard a dog bark. That made us happy. It was one of McClatchey's sheep camps and they had just come from Harlowton.

Thursday morning I got Dick and found that the saddle had hurt him so that it would be cruelty to get on him, so led him over to a cabin and sheds, unsaddled him and turned him loose. . . . We went over to exposures on Mr. Middecombie's where there were large bones in abundance. I think this will be a good field to work. . . . I was anxious to get home so I started afoot, at almost dark, and Mr. Willis rode horseback. The only fear was that it would rain and that we would not get across the Mud Cr. bridge before dark. It did not rain and we made the bridge, though it was dark. The last mile or two Mr. Willis insisted on my riding. We got to camp a little after 10 p.m.

August 22, 1901

I began work for Dr. Farr. Started for the Lake Basin. Mr. Spear was with me. The first night we camped at a sheep camp below Flanigan's house.

We established camp here on a little stream. The 26th I started for Columbus to get my father and get some provisions. I went to Mr. Irwin's. Mrs. Irwin is she that used to be Minnie Beard. I had not seen her since she, Mrs. Evans, Nettie and I, drove from Brookings S. Dakota to Minn. to visit her folks. Cora is married and has two children.

They gave me a bed and I got up early in the morning and went to meet the train. The train stopped and he (Earl's father) got off. He was not looking very robust, and told me he had spent a horrible night. He said walking eased him. We walked until light then found Mr. and Mrs. Irwin and called the Dr. He made him some tea and used the catheter and he was immensely relieved. . . . He is now well and hearty. He does lots of work and helps dig out the dinosaur and I would rather have him than anyone else I know.

September 24, 1901 - Camp in Lake Basin on A. Whitney's ranch.

There has been considerable change since I wrote last. The Princeton party has all returned except myself; all who are going east. The party broke camp August 15. . . . Willis went to California. Spear stayed with me until September. The rest went to the park. I took the specimens to Harlowton and shipped them. They paid my salary to the 16th and $6.00 for taking specimens to Harlowton. I sent $50 to Father to come out here. . . .

Before going away Mr. Farr had asked me to get the specimens I had found when Mr. Willis and I were on our prospecting trip. He said he would give me the same as I had been getting, $100 per month and provisions furnished.

The 19 of August I went to collect leaves. I did not go to the place I found the leaves last year, but struck for the stratum further north along the bluffs. Saw a bare side hill, and as I always do, examined the shale for leaves etc. I supposed these beds were Laramie or Ft. Union (formations).

While going along the hill and slowly ascending, I saw a piece of fossil tooth. I looked around and found another piece and then a smaller one that nearly made the tooth complete. "That looks like a mammal tooth," thought I. Is it possible that there is any reptile that has a tooth like that? I was not sufficiently expert as a paleontologist to be certain, but on looking around I found a premolar; that settled the thing for ever. I had found mammal remains. I examined the shale carefully and systematically and found more fragments. I had found mammals in strata that I thought was way back in the Cretaceous or borderline between the Cretaceous and Tertiary. It was one of those discoveries that makes a paleontologist happy.

This seemingly insignificant jaw and teeth, found August 19th while looking for leaves, turned out to be an important discovery. Professor Scott termed it, "An epoch-making discovery." The geological age of the Fort Union formation in which it was found had been in dispute some forty years. Leaves and fresh water shells had been found in abundance but they did not definitely determine its age. My father's discovery fixed the age geologically, correlating it with the early Tertiary deposits of New Mexico. This deposit was laid down during the transition period, between the Age of Reptiles and the Age of Mammals, during which time the early forms of mammals were starting to appear while the dinosaurs were not yet extinct.

September 25, 1901

Today we went to the dinosaur and worked at it several hours. Did not get out yesterday on account of the storm. Did not go out early this morning as the buckskin mare evaded capture. My horses now are Billy, the sorrel cayuse that I bought last year in Bonner for $20. He is sorrel with a white strip in his face and white below the knees behind. He is very good except on the pull. He does not understand that very well and gives up if it seems discouraging. He is subject to fits of stampeding and does not use any great amount of judgment when he gets scared. The other horse is a mare, a buckskin with a white strip face. We bought her of W.A. Clark in Big Timber with 4 other broncos. I took her for $50. She is a pretty good mare, larger and stronger than Billy. Her name is Snake. I have the light wagon I bought of Todd Rogers for $25 during the latter part of the last century, about 1896. It has done good service. The harness is the one I bought of Mr. J. Freeman in 1897. It is a pretty good harness yet. Have the small miners tent and the cot and saddle I bought of Sears Roebuck and Co. about the same time.

We are camped in a favorable locality. Good wood, water and feed. We cook and get our meals in a log cabin. This was a winter sheep camp and was dirty and smelled badly on account of the former occupants, the mountain rats, who are anything but tidy. Mr. Willis, when he was here, shot one underneath the table with his revolver, and a few days ago I stabbed one with my double barrel shotgun as he was resting up beside one of the boxes of the dinosaur by the stove as it was curing after being bound up in cloth and plaster.

We are nearly 30 miles from any market and do not often see a person, except ourselves, unless we go over to Whitney's (3 or 4 miles).

After my father and grandfather had removed the dinosaur bones and taken them to Big Timber for shipment to Princeton, they proceeded westward up the Musselshell River to its head. They left Mr. McClatchey's, who had been very kind and would take very little pay, the 9th of October. After leaving the Musselshell River they circled to the north of the Crazy Mountains then headed in a southerly direction for Belgrade collecting fossils along the way. After spending a few days in the vicinity of Logan and Three Forks they proceeded to the lower Madison Valley where the team was left with Mr. Hutchinson for the winter.

Before returning to Princeton Father made another visit to the University at Missoula where part of his collection, which was still stored there, was shipped to Princeton. On his return he stopped in Dillon where my mother was attending the State Normal College. The visit in Dillon is related from his diary.

> November 7, 1901 - on train going east in North Dakota.
>
> When I arrived in Dillon I went to the hotel across the street. Went into the waiting room where she said she would be. The room was not well lighted and when she saw me, she ran and threw her arms around me, and then we greeted each other as we should. We walked together for a while. The next day I went and called on her at about 8:30 in the morning and we went up to the Normal School. She introduced me to President Monroe. I attended the classes and

spent the time very pleasantly. We went away together and returned in the afternoon. She got excused from one class and got out early. I hired a horse and buggy and we took a ride across the river. Had a nice horse and buggy. In the evening we visited in the parlor of the house where she boarded. Had a splendid ride and a pleasant visit.

⌘ ⌘ ⌘

CONTRIBUTIONS TO SCIENCE:
A Position with Carnegie

Father returned to Princeton November 14, 1901 after spending four days with his father and sister in Minnesota. He immediately went to work preparing his fossils, including the specimens which had been stored at the University of Montana, for study and description. With several new discoveries he decided it would be best for the benefit of science to write a preliminary paper revealing his discoveries rather than have them revealed through singly-published scientific papers which might take years. He set about this task with considerable enthusiasm.

> December 23, 1901
> At the museum working as usual. Am preparing a preliminary paper giving a brief diagnosis of all my new species. Some have been in my collection 6 or 8 years and are yet unknown to science.
>
> Tuesday, December 24, 1901
> Have been studying this evening, trying to find to what genus a lower jaw I got in the Madison Valley years ago, belongs. Have described it this evening.
> Some time ago I started to write something for the newspaper. "The History of a Dinosaur", with sketches etc. but I am so anxious to get out the paper I spend nearly all my time on that. I think I have something like 30 new species.
> I am boarding myself now most of the time. Do not know how long I can stand it.

Christmas day Father went to Trenton and the next day he had a surprise when trying to reconstruct a fossil jaw, the center of which he supposed was missing.

> Friday, December 27, 1901
> Christmas I got up feeling in a decidedly good working mood. Worked a little while and got decidedly out of the mood. . . I decided I would take a ride to Trenton. I found there was an opera, a matinee, so bought a ticket. There were lots of actors and the costumes were good enough but did not like the opera. There was one good character, however, the lunatic who was going to kill Mr. Bronson.
> Yesterday I received a couple of presents from cousin Laura and a sweet picture from Pearl.
> Have been at work all day today preparing material. Have an unknown animal I think. It is in hard matrix.
> Yesterday I picked up some pieces of a dog jaw that I had made some effort before to get together, and to my surprise there were two pieces that fitted together, but the lower jaw was so surprisingly short that I thought there were pieces gone. It is evidently a new species with a very short jaw and face, with teeth condensed in size and number and crowded, the premolars, some of them, being placed diagonally.

In January 1902 the little fossil mammal from the Musselshell arrived. This aroused considerable interest and consumed much of his time.

> Thursday, January 9, 1902
> Day before yesterday the box that contained the Ft. Union and lower White River fossils came. I opened it and unpacked most of the things that evening. The next morning when I went down to work Professor Scott and Dr. Farr were looking at the Lower Eocene things. Professor Scott said they were very interesting. He wishes me to publish notice of finding the fossils and to send a set of fossil leaves to the U.S.

Geological Survey for determination so we can fix the thing and, if possible, settle the age of the Ft. Union beds.

Dr. Ortman was interested in the crustacea. The star fish is new, only two of them having been found in America near this age. He wishes me to write this up also, naming the star fish. I do wish I had money to go to Montana and do as I wish.

Dr. Ortman was a well-known invertebrate paleontologist at Princeton who later went to the Carnegie Museum. The starfish was found when my father and grandfather were on their way from the Musselshell to Logan, Montana the fall before.

February 4, 1902

After Professor Scott requested me to send a notice of discovery of the mammals from Montana to "Science", I went to work and got immensely interested. Got book after book and read of the discoveries concerning the Laramie and Ft. Union Beds. Went to New York to compare and identify the fossils, and to Philadelphia to look up literature, and then came back and read again. Came to conclusion after conclusion independently and found that other men had suggested the same ideas before. At last I got about 10 pages of 'hardly' wrought material, showed it to Professor Scott and we concluded it best to send only a small part to Science. That I sent several days ago.

March 13, 1902

Dr. Mathews was here Monday I think it was. He examined my collection. Mr. Brown of the American Museum was with him. Had an interesting time with them, especially Dr. M... Some things are in the line in which he is working. He wants me to go to Montana for the American Museum next summer. Said he would talk to Professor Osborne. As Hatcher had spoken to me last year I thought I ought to write to him before making a bargain with American Museum. Today I received a letter in which he wished to know if I would engage with the Carnegie Museum as Assistant Curator. I have spoken to Professor Scott, Farr and my friend Mr. Miller and they all think it is the thing for me. I would rather be a professor in a university in the west, perhaps, or a state geologist, yet this will allow me to divide my time more exclusively to my favorite work. I had been a little blue about my immediate future for the last few days. This makes things look brighter or will I get it?

Am taking Vertebrate Paleontology by Scott, chemistry by McKay, studying embryology and osteology and painting during my leisure time. So I manage to keep busy and get tired sometimes.

They have been having a terrible time at Dr. Farr's. Dear little Jamie is dead. He had the Scarlet Fever. The baby did not have it. Vernon was very sick but is slowly improving.

Sometime in May Father went to work for Mr. Hatcher, veteran fossil collector, and Curator of Paleontology at the Carnegie Museum.

May 30, 1902 - Pittsburgh, Pennsylvania
I had two offers - 3 including Farr's offer to go out in the field this summer. Only one was an offer of permanent work. That was a Carnegie offer and I accepted it, and am here now.

He was not at the Carnegie Museum long until he was sent to Montana to collect more fossils. He arrived sometime in June. As will be seen, his contemporaries were also in the field retracing his footsteps, presumably hoping to pick up what he had passed by. He, however, on a hunch picked a new area which he had passed by before and it turned out to be very productive.

Arrived in Helena on the Great Northern Tuesday afternoon....

In Helena I went to all the livery stables to look for a riding horse but only one had any to sell. He was not worth much but they asked $45 for him. . . . I did not know how gentle he would be to ride. Do not like Helena so did not stay long.

I had on several occasions noticed the exposures of the Tertiary deposits near Winston so took the train for that station. Arrived there between 3 and 4 p.m. Went over to the hotel and engaged a room, got a prospectors pick and started for the exposures.

Ascending a hill, covered principally with igneous material, I was delighted to see grand exposures of what I was looking for and quite extensive. I was full of courage and the highest hopes. . . .I got caught in the rain, hurried to the Missouri River bottom, found a house, but the lady said they had no bed. She directed me to a neighbor, Mr. Beattie, who had a nice large house. Mr. Beattie is a hospitable friendly westerner.

The area mentioned around Winston is about twenty miles southeast of Helena and is the area between Highway 287 and the Missouri River to the northeast. After staying overnight with Mr. Beattie, Father went to Winston and picked up his grip, luggage etc. He continued to stay at Mr. Beattie's until the night of the 17th when he stayed at Mr. Cristie's who had a ranch on the north side of the river. The next day he bought a horse, bridle and saddle from Mr. Cristie for $30.00. He named him Archippus (Greek for "ancient horse") and commented that the horse was old and slow but very gentle.

The beautiful exposures on the south side of the river, around the Beattie ranch, produced nothing but some plant impressions but bones were soon found when he started prospecting the area northeast of the river near the base of the Big Belt Mountains. After searching this country for a few days, and after seeing many bones and jaws sticking out of the rocks he decided to recruit some help and collect some of the material as it looked very promising. This area turned out to be very rich, especially near Canyon Ferry where there was a dam across the Missouri River and a power plant that generated power for Helena.

After deciding that no more prospecting was necessary in the vicinity of Canyon Ferry Father rode horseback to Townsend then to Whitehall, prospecting and collecting along the way before going to the Ruby Valley where he spent the 4th of July at the Goetschius ranch. He returned to Whitehall July 6th and from there went to the Lower Madison Valley where he had left his team the previous fall. His plan was to leave there for Canyon Ferry as soon as practical. Before leaving, however, he revisited his first prospecting grounds where he had returned so many times. He was quite certain its fossils were well-depleted but was surprised when he found a ledge of different material in the sandstone back of Big Round Top that contained petrified wood, vegetable matter and bone. This he believed was the site of an ancient water hole, used by the prehistoric Miocene mammals. From it a quantity of fine material was taken, some of it being even better than his original discoveries.

This caused considerable delay so that he did not arrive at Canyon Ferry until August 1st. He took his team and wagon, Archippus and camp outfit with him and hired as an assistant a Mr. Roberts from Logan, and his old friend Mr. Hutchinson.

> Sunday, August 3, 1902 - Canyon Ferry
>
> We arrived here day before yesterday. Started from Logan, July 30, a little after 11:00 A.M. We met Dr. Mathews and party (Dr. Mathews of the American Museum, SAB) between Logan and Three Forks. They had been over to Pipestone. Said they found some little things. I had visited Mathews in camp on the Madison and especially the last time, he did not seem friendly and it hurt me, but he was his old self when we met last.
>
> Yesterday we saw portions of 15 or 20 skulls here, all oreodonts.
>
> August 11, 1902 - at Stubbs Ferry
>
> We came over from Canyon Ferry yesterday. Drove my team as we wanted to look over the country and felt more like spending Sunday that way than staying in camp. Ate our lunch and when I went to get the horses, to hobble them, they were

very nervous and jumped, jerked me down and got away. We stopped here last night after walking 10 or 12 miles or more. They hurt my knee and I am lame. It tired me to walk and it seems that I could not have gone so very much farther.

Mr. Hatcher spent Friday and Saturday with us and we got out 7 or 8 skulls, most of which we had dug around. I learned lots about collecting and did not like to see him go. He gave me a good deal of good advice. I think I can do better now.

This seems a fitting place to digress from the diaries to tell of an incident which happened some twenty-seven years later in the same locality. My father had just finished a job, as expert geologist for a large Gilsonite Company in Utah, during the summer of 1929. I had been working with him a couple of months in Salt Lake City while finishing the job. He had mentioned many times his desire to return to his old collecting grounds which he had not visited since his early collecting days. When the work in Utah neared completion he suggested such a trip, offering to pay the expenses. The year before, while working at the Goetschius Ranch in Alder, I had purchased a new 1928 Ford Roadster, one of the first to enter the Ruby Valley. (Father never learned to drive but loved to travel by car, trusting anybody's driving.) I was agreeable since I knew we would end up in Montana where my mother was at that time. Also, I had a girl friend there that I was quite anxious to see.

We started from Salt Lake City and took a leisurely trip through Wyoming and into Eastern Montana then westerly across Montana arriving at Canyon Ferry after about ten days. We collected fossils along the way finding two beautiful skulls, an early horse and a rhinoceros in the Three Forks area. Father was in his glory, having the first opportunity to make a private collection since his early days in Montana, and having me to drive him around, help dig out the fossils and lug them to the car.

Up until this time I had enjoyed the trip quite well, but now, it being only a short drive to the place where I might see my girl friend, I was beginning to get a little impatient. Not so with my father. He started systematically searching the white bluffs as he had done years ago. The second day, having found nothing of value, I practically lost all interest in fossil collecting and was trying to amuse myself by searching for arrowheads. High above me at the foot of a huge cliff he called to me. I climbed the long talus slope in the hot sun expecting to see some rare specimen but to my utter surprise and disgust he was feverishly pecking away at the bare cliff with his prospecting pick, with nothing showing but soft white sandstone. All he said was, "There might be bone here". I didn't even offer to help but just sat in amazement and watched wondering how long this would go on. Nothing more was said for some twenty minutes. He worked away and after digging a hole in the cliff about four inches deep, to my utter surprise he struck bone. I got up then and went to work. Before the day was over we took out a beautifully preserved and absolutely perfect skull and lower jaw of an oreodont. (The oreodonta is from Order Artiodactyla – two-toed ungulates, Suborder Suiformes, not from the tylopoda, which are camel relatives, SAB) I asked Father what made him think there might be bone there and he showed me there was a very slight discoloration of the rock for some six inches around the bone, a fact he had discovered years ago.

> August 16, 1902 - at Canyon Ferry
> In the John Day beds where we are now at work there is more evidence of stratification. The lowest beds exposed on Fossil Hill are sandy. . . . The fossils are surrounded by a lighter colored rock. Most all bones that occur outside this are broken or rotten.

The remainder of the summer was spent at Canyon Ferry and although there is little in the diary concerning the activities, it is evident a considerable amount of choice material was taken out. In an article published in The Guide to Nature Magazine, April 1908, he stated that there were between twenty and thirty boxes of precious fossils sent east from this summer's work.

⌘ ⌘ ⌘

23

EXPEDITION OF 1903:
A Troubled Love Affair

Following his summer at Canyon Ferry Father again visited the Ruby Valley and my mother and the Goetschius family before returning to Pittsburgh. Although the discoveries made during the summer were very gratifying, matters concerning his love, which were of greatest importance, were not going well at all. My father, seventeen years older than my mother, was attracted to her even when he was her eighth grade teacher, in a little log school house, the first year he entered the Ruby Valley. He had always visited the Goetschius ranch when in Montana and gradually a romance had developed, although according to his diaries he had tried to prevent it because of the difference in their ages.

My mother was a native of the Ruby Valley. She had known the hardships of pioneer living but had also seen better times when her family had prospered. Her father, John Goetschius, a native of Pennsylvania, had crossed the plains with wagon trains in 1870 and took up a homestead in the Ruby Valley where he developed a ranch. Soon after homesteading he married Charlotte Whitmore, a young woman who had crossed the plains with her family in a wagon train and entered the Ruby Valley about the same time. Although Virginia City, a short distance away, was still a booming mining camp, John Goetschius never participated in mining. He went immediately to raising hay, grain and cattle making more, perhaps, than did the average miner. His produce was sold almost entirely to the miners in Virginia City.

Pearl was the eldest in a family of five children. She was born in the little two-room log cabin her father first built. When she was a small girl and her father was away, she remembered a band of Indians coming and demanding food. When refused they went to the pedal-operated grindstone in the yard where they sharpened their knives and kept laughing. Pearl's mother finally gave the Indians practically all the food in the house to get them to leave.

By 1900 a new house had been built on an adjoining piece of land which had been purchased by her father. The ranch was one of the nicest ranches of the valley.

When my father's diary continues, after an interruption, he had returned to Pittsburgh and the Carnegie Museum. The last visit with my mother had ended in a broken love affair. What had been contemplated as a pleasant buggy ride ended in separation and could have been a tragedy. What happened is gradually revealed as the diary continues.

> November 15, 1902 - Pittsburgh, Pennsylvania
>
> At last I have written Pearl. It seems I could not help it. I had a conflict from the time I left there between love and pride, between tenderness and bitterness. I have every day my feelings of deep regret and sorrow and fears for the future, wounded pride and longing for what is not.
>
> I went down town this evening. Among other things I purchased Renan's, "Life of Jesus", Farrar's, "Seekers After God", Carlyle's "Heroes and Hero Worship" and Zola's "Dr. Pascal". I have been reading from the two first this evening and it has done me good . . . I have been inspired to rise above these little mean things like jealousy and resentment. I hope it is my privilege to live in a higher atmosphere. If I can rise above it I may find the love my soul craves, even if Pearl marries someone else. I may never love any individual as I have her. I don't expect it. I do not think I will ever marry though, of course, I cannot tell. I want my better days to be full of work and to do something to encourage and help my suffering fellow men. May my inspiration continue.
>
> November 16, 1902 - Pittsburgh
>
> Went to Bible Class at the Presbyterian Church, where I attended a week ago tonight. From there I went to the

Unitarian Church at Craig St. ... The minister gave me a view of Jesus so different from the ideas usually held by Christians. ... Christ suffered and died wholly as he had lived, but the idea that he was the Son of God more than other great souls, would, it seems to me, detract from the significance of his life.

I realize that my struggles are not over yet by any means. I have much to suffer on account of this human love of an individual. Sometimes in my more sublime moments I feel that I can rise above it but again I know that my feelings are so profoundly connected with this love. Oh, if I could love as I have her, someone who could love as she seemed to, it seems to me that I could rise to heights I have never reached. How sweet to live in sympathy of kindred spirits, to work and live and love together in perfect sympathy. ... It seemed once within my grasp but now it is gone. She who loved me loves me no more. She who sympathized with me has turned against me.

November 17, 1902

These are days of change for me. At the age of 40 when men's minds are usually tired, the fountains of the great deep in me are stirred up and I feel that after this I will be worse or better. ... To despair and willingly give up I cannot, on account of my folks. I must be a man. I sometimes think death is preferable to what I will suffer but it is not for me to choose. I must make the best of it and live for others, not myself.

I have for a time loved someone. ... I perhaps could mention 30 or 40 that have affected me for a longer or shorter time. Not until I met the last one did it endure. It seemed that something held us together. I suppose it was my great love. She answered every objection. I almost forgot my age and seemed almost as young as she. I would, if she had continued as she was, have loved her until my life closed. But when our highest hopes were on the eve of consummation, her heart turned from me. She went into the company of others, which I wanted her to do, though I did not want her to be almost constantly in the company of one and then let her heart be so easily led from me. Perhaps I should not say easily, I have not seen the gentleman and he may be more charming than I am at this time willing to admit.

But after all I am somewhat to blame and I may realize it now. She was honest, and told me about going with this man. . . . I did not like that I confess. I wrote no reproof but maybe I did not write so tenderly. . . . Until that last miserable time that we met it was all kindness. When I met her last I soon saw the change, that she did not enjoy my company, that others had more attractions for her. She treated me shabbily, paid little attention to me when others were around, did not invite me to stay. What a different girl she appeared from what she had ever seemed before. We went riding and there were no kind words but coldness. We had an accident and she blamed me though no human being could have stopped the horses. But so strong and long-enduring had been my love, so sacred to me, I wrote to her after all this. If ever she should repent of what she has done I could forgive her. . . . It may be that some time she will realize that my love was not a thing to be lightly flung away.

The gentleman mentioned was the son of one of the most prosperous ranchers in the valley who lived a short distance from the Goetschius ranch. He had a nice carriage and drove fine horses. He and my mother were near the same age and had been friends since childhood. They both loved to dance and my father had never acquired any great liking for this sort of entertainment.

The accident with the horses was a well-known incident in the valley. My mother's younger brother, Grover, remembered it well. My father had rented a team and buggy from the livery stable in nearby Alder and in the evening they went to visit the Cahill's who lived at the upper end of the valley, a distance of four or five miles and stayed quite late. When returning home, suddenly and without warning the neck yoke came loose and the tongue of the buggy went down at a place where there was a curve in the road, a steep hill on one side and a large irrigation ditch on the other. When the tongue went down the team stampeded turning the buggy over in the ditch which was filled with mud and water. Fortunately neither were hurt but both being dressed in their best clothes must have been a sight to behold as they crawled out of the muddy ditch. The team was left at a nearby ranch and they walked the rest of the way home. Grover remembers hearing them come in late at night. The next day he noticed the beautiful ostrich plume from my mother's hat, which was the style in those days, lying in the mud by the front gate. It remained there for several days.

It was believed by some that a couple of young boys at the Cahill place might have been bribed, by someone not particularly interested in their going together, to fix the neck yoke so it would cause a mishap. The anguish between the two left both bitter but my father turned to his scientific interests with renewed vigor while trying to recover from his disillusionment.

> November 18, 1902
>
> Cheer up Douglass, don't let your life be affected by one, yet it is almost always one more than all others, perhaps, that affects our lives. One thing, I am well. My health has not been better for years. My prospects are, in a way, brighter. At last there is a prospect that after a while I will get a fair salary, that is for a scientific man. I have now a little ahead...a little but that is better than I have had before. It has been a life-long struggle with poverty, and poverty has, so far, gained the day.
>
> November 20, 1902
>
> Professor Osborne (Osborn, SAB) was at the museum today. I think very highly of him and honor him much for giving himself to science for the pure love of it, when he has all other that wealth can buy.
>
> The carload of fossils I started from Billings came today. We have unloaded and stored them away. Mr. Utterback's

and Mr. Peterson's (Olaf A. Peterson, SAB) collections were in the same car.

During the ensuing months in Pittsburgh a great deal of Father's spare time was spent in reading and delving deeper into the problems that still bothered, yet deeply fascinated him, i.e. the conflict between science and religion. Churches of various denominations were attended regularly and lectures at Carnegie Hall were often attended in the evenings. In January he received a letter from my mother.

> Monday, January 26, 1903
> Last Friday I received a letter from Pearl, the first one since our separation. She, poor girl, has been more miserable and unhappy, I fear than I. It touches my heart with pity.

> February 22, 1903
> This afternoon I walked down to the Washington Streetcar line and beyond. The scenery is quite picturesque. I thought of the possibility of living over there and had some strange feelings about the one who was to have been mine, but it seems never to be. And yet in the strange complexity of things a dim hope rises and it brings up the spirit. Above the firm foundation of the hills, above the wooded glens, above what we call the realities of life, and in spite of the hard things we call facts, one feels that far off, somewhere, somehow, good and truth and love will conquer and there is peace.
> Attended the lecture at Carnegie Hall this evening. It was on Socrates and was extremely interesting and instructive.

> Sunday March 1, 1903
> My reading Spencer's "Principals of Biology" inspires me and I long to read "The Origin of Species", "The Descent of Man", about electricity, astronomy, and in fact nearly all the branches of science. . . . I am getting where I do not fear, at

least as I used to, to go to the greatest depth one can reach in science.

March 4, 1903

Mr. Hatcher and I had a talk about next summer's work. He thinks he may not be able to send me out next summer. This is a disappointment. There is one advantage, if I stay here I can continue the studies I have started.

March 8, 1903

Today brought quite an experience to me. By the solicitation of others I was brought to consent to prepare something for the Sunday School. But I had serious misgivings about standing before the intelligent people there and trying to express my thoughts. . . . At first only 3 or 4 assembled in the room where the S.S. was to be and I thought it would be pretty easy. Finally the bell was rung and they came in and filled up the chairs and I must take the consequences. My cowardice that prompted me to stay was greater than the cowardice that prompted me to run so I stayed. I got up and talked quite freely though my legs felt in quite a tremor but by leaning up against a piano or chair, I managed to stand. All seemed interested, however, and there was quite a discussion after and several spoke very highly of the effort. I hope I can do better next time, and am glad I had the experience.

Yesterday forenoon I received a letter from Pearl. I did not dare read it until I came home. Mr. Hatcher saw me slip it in my pocket unread and joked with me about it. I dreaded to read it. It showed me that, try as I had to subdue it, I still hoped. But she has not forgotten me. It seems she did not get the last letter I wrote after the death of her father. She has been worrying about me. She wants to give me a little present, a ring to remember old times by, if I will accept it. I should prize it far above rubies. Although I have changed, in spite of all that has happened, my heart still clings to her. She seems the only real girl to me.

May 12, 1903 - Pittsburgh

I am passing the time quite calming. Have not had any of those deep sorrows lately. For some reason the knowledge that I am something to the girl I loved and from whose influence I have not yet escaped, seems to make my sorrows less and I feel more free.

Prospects of seeing Pearl in the near future filled Father with fear, yet hopes, that a miracle might take place. Previously doubtful plans for a trip west were finally arranged. There is no mention of preparing for the trip but in later entries he related that he left Pittsburgh May 19th and after visiting the Field Museum in Chicago and his family in Medford, Minnesota he arrived in Logan, Montana June 2nd.

He engaged his old friend, George Darlington, of the lower Madison Valley to help him and they started the summer's expedition June 4th. The summer's work was not designated purely as a fossil-collecting trip but rather as a study of the lake beds to gather more information regarding their extent and climatic conditions during Tertiary times. They arrived in Virginia City June 25th after spending some time around Three Forks and the upper part of the Madison Valley.

July 1, 1903

Stopped in Virginia City two nights, June 25 & 26. Saw many old friends and had a good time. As I approached the Ruby Valley it was with some feelings that were not very pleasant. It was the scene of "almost" a tragedy and one of the saddest events of my life. . . . I did not know that I could approach the little valley again, where I had been so happy. I sometimes felt I could not go see her. . . . The death of her father was an awful blow. No matter how she had treated me could I refuse to go and see her or try to comfort her in a dark hour? Perhaps if it had not been for this I could not have had courage to go and see her.

We went down to Mr. Taylor's and camped on his ranch. They were as friendly as ever and we had a good time there.

I took supper there and then took a horse and rode down to Mrs. Goetschius's. . . . I did not know but seeing her again might help me. Perhaps that love would depart and trouble me no more. But I never talked to her (Pearl) when she seemed dearer and how I wished that I could have her love. We talked on subjects that were near to our hearts.

Then we talked plainly of sadder things that brought an inevitable pain to my heart. I think that she and her lover will yet be married. I understood her to say that he had been told of her feelings. I believe he still loves her.

My father left the Ruby Valley July 2nd and traveled to the north end of the McCartney Mountains a short distance south of Melrose. He had noticed an exposure of lake beds on the south end of that range some years before but could not reach them because of the Big Hole River. He now tried to locate it from the north end of the range. Because of a steep hill that could not be pulled with the wagon he went on horseback. After a considerable search he finally found the exposure and though it was a small area it contained some valuable material. As it was a horizon in the lake beds where little had been found before, the discovery was important. The camp was moved to the Beacher ranch where they stayed while removing the fossils.

July 7, 1903

Yesterday morning we pulled up camp. George got Mr. Browne to help pull us up the steep hill this side of Mr. Browne's. I think it is the steepest long hill I ever tried to get up with a wagon. It made 4 horses get down and pull about all they could to get our light load up.

When we got to Mr. Beacher's he came out to the gate and said he had a proposal to make. He said we could go down to a cabin about a half mile that was used for a school house, and use the cabin. He said there was a stove in it. It was very kind of him. I had never seen him before. He told us to drive over and then go back and have dinner. We gladly did so.

In the afternoon we went to the fossil beds. Found two or three more skulls.

Sunday, July 12, 1903

I am sitting on my cot in the log cabin that has recently been used for a school house. The floor is made of 10 in. unmatched boards worn so that the knots are prominent. The cabin is built of fairly large-sized logs and overhead is cheese cloth sewed with large twine through the middle and held up by strips of wood. Near the south corner is a small cupboard, made of a pancake flour box with one shelf. ... All the walls have been whitewashed.

Looking out the door to the southward is first a weedy back door yard then a little plowed strip. Beyond is a strip of alfalfa, apparently recently sown, then a worn fence, then a strip of brush, then the dark green of the cottonwoods, and beyond that rolling grassy hills and a ridge of dark looking rock.

We have been up to Beacher. Took dinner there.

The mosquitoes, gnats and flies are extremely bad now. The mosquitoes hardly let up day or night.

Since I saw Pearl last I am still distracted and my mind unstable. Outwardly it seems a one-sided question. ... Mr. P. is younger, is prosperous. ... Yet in the face of all this I cannot give her up until I have to. I sometimes think I am a fool.

Sunday, July 19, 1903

Hot. Mosquitoes and flies are terrible now.

I think we have nearly finished collecting in the place where we have found so many things.

Today I had callers, four young ladies and an English gentleman. The young ladies are school teachers. Two of them I met when I was in Dillon a couple of years ago. One was Pearl's room mate. I would lose my heart again if I were with her much I fear. Pearl was mentioned and it made me sad.

There are three of the Beachers here, Charles, George and Frank. The latter two are in partnership and live together. They are both married - have wives much younger than themselves and have, in all, six children, all small. I think they lived here a long time batching. They married German women. There are, I think, only two other families on this side of the river. The Beachers have two good log houses, a stone house, milk house and cellar combined, a shop, a hen house and the best pig house I ever saw. They keep lots of cows, hens, pigs, and horses. They raise grain, alfalfa and wild hay. I like them very much and they are very kind and generous to us making us welcome and accommodating us in every way they can. They have an old man who is the gardener and several hired men to look after the horses. They have been gathering some in lately that they have sold.

August 12, 1903

Have quit fossil hunting for a while. Mr. Darlington started with a load of 7 boxes for Laurin yesterday morning. Mr. Morgan telegraphed that he would meet me in Butte the 13th. I made arrangements to meet him there.

Mr. Darlington is going home until we get this business straightened up. I am at the Southern Hotel this morning.

Mr. Morgan, my father's old Pittsburgh acquaintance who owned and operated a hardware store, was interested in the possibilities of gold mining in Alder Gulch, which my father had brought to his attention. Mr. Morgan decided to see the area himself with the idea of possibly investing some money there. He was delayed and did not arrive until the 19th, at which time they proceeded to Virginia City where they rented a team and buggy. They were particularly interested in the gulch from Virginia City to its source, a distance of about five miles. Although the rich placer gravel that had continued to the foot of Old Baldy Mountain and abruptly ended within a distance of a few feet, had been mined out by 1903 there was still considerable activity on the load veins. For years a post with a sign that read, "The end of the God Dam Pay" stood at this point, probably signifying the disgust of some miner whose dreams of riches had suddenly ended. After examining the area quite thoroughly, it was decided by Mr. Morgan and my father not to invest any capital.

In 1905 an article by my father, "The Source of Placer Gold in Alder Gulch", was published in Mines and Minerals Magazine. After many years of study, thinking all the time that a rich vein might be found, his final opinion as presented in this article was that the gold came from a number of fairly low-grade veins which more or less paralleled the gulch instead of the gulch crossing them at right angles as is usually the case. Thus, in a course of several miles, many times the usual amount of vein material had been eroded into the gulch, enriching it immensely. Though much money has been spent in the gulch since 1905 by those prospecting for gold, the conclusions of this article have not been disproved to date.

After completing the business in Virginia City Mr. Morgan returned east and my father proceeded to Divide, a little town on the railroad a short distance north of Melrose, where he joined Mr. Darlington.

> Thursday, August 27, 1903 - at hotel in Divide.
> Arrived here this morning. Mr. Darlington came on the evening train.
> When I get time I want to write up more fully my Virginia City experience. One thing nearest my heart I wish to write now. While at Mr. Beacher's I made up my mind I would write more to Pearl to try to cheer her up. I suppose there was a selfish wish in it too. Well I wrote two quite long letters. I received no reply. Felt that she and B. had seen each other, made up and I was left. I had been expecting it. . . . I was quite near her for 10 days while at Virginia City but did not think it best to go and see her. Did not expect to see her again. Well, the night before we left Virginia City I received a letter that had been forwarded from Melrose. It was a good confiding letter. It showed that she had far from forgotten me and it seemed more and more that we might yet be reconciled. I, of course, concluded to go and see her, and it

perhaps was one of the happiest days of my life. She was so kind, so sweet, so seemingly affectionate, that a whole heavy load rolled from me. Bitterness and despair rolled away and left me happy again. But I must retire as I am weary.

With the happy reunion of the Goetschius ranch, after the long interrupted love affair, a different outlook seemed to dawn on the horizon. It now seemed the love affair could be repaired and marriage was quite probable, as will be seen in future diary entries.

After doing some prospecting in the area of Divide, my father and Mr. Darlington proceeded up the canyon of the Big Hole River with the idea of exploring the Big Hole Basin and then going to Dillon by way of Bannock. After traveling through some twenty-five miles of canyon the basin comes into view. It is fifty miles long and ten miles wide and is completely rimmed by snow-capped mountains. There are stacks of wild hay, for which it is famous, almost as far as the eye can see. Dewey, which is mentioned, is in the canyon and Wisdom is the main town in the basin.

>September 3, 1903 - near Fox, Montana
>Since writing last we made quite a journey...Was glad to get into new country... Dewey is an old mining camp badly rundown. Wisdom was a surprise to me. There are two nice general merchandise stores, two or three hotels, meat market, barbershop, laundry, drug store. It is a little place all in a bunch and things look new. It is one of the nicest little burgs I have seen anywhere.
>A little distance this side of Wisdom there was a man lying on the ground, hurt by a team running away. A man caught the team and went for the Dr. We stayed until Dr. McNevin came and they took the wounded man on home.
>This is a great hay country. Nothing but hay. Of course they raise a lot of cattle and some horses.

A few days were spent on Grasshopper Creek above the town of Bannock where some good fossil material was found. Bannock was the first capital of Montana and the site of the first gold discovery, of any consequence, in Montana. It was here that Henry Plummer, the famed leader of the road agents, was hung. By 1903 it was a ghost town, much as it is today.

    The collecting trip was terminated in Dillon. After a visit to the Ruby Valley Father returned to Pittsburgh to resume his writing and to make plans for a home and his contemplated marriage, which he hoped might take place in the not-too-distant future.

⌘ ⌘ ⌘

PITTSBURGH, PENNSYLVANIA:
Preparing a Home

My father's attitude and outlook for the future was now quite different from that of the previous fall. Instead of heartache, melancholy and pessimism, which had previously engulfed him, he was now cheerful. Although there had been no definite time set for marriage he did not seem to think it was necessary to wait any longer to think about owning a home.

> October 14, 1903 - Pittsburgh
> Arrived in Pittsburgh last Sunday morning on the 2:45 a.m. train. Felt somewhat weary but as it was so late concluded not to go to bed. Went over to Jerry's Restaurant to kill time and get a lunch. The waitress pitched a side order of potatoes into my pocket. I did not think it very funny at first, but she laughed and all in the house laughed and I soon began to see the humorous side of the waitress and laughed too.
>
> October 15, 1903
> I am very happy of late. Have my troubles but it seems that there is only one who can give me content happiness and peace. It is interesting to try to advance my way in the world for her sake, to have her encourage me though she is far away.
>
> Saturday, October 24, 1903
> Am now working out a skull, a good one, that I got at Canyon Ferry.

Have not heard from Pearl since I came here and have begun to be disappointed the last two or three days but if she is well and happy it is all right. She told me not to be disappointed if she did not write much. The poor girl has so much work to do. She is teaching near home and boarding at home and she cannot sit down and see her mother work.

Wednesday, October 28, 1903
This is my birthday and I am 41 years old. I don't know that I need to be melancholy and stop to wail about the shortness of life and the approach of old age. My health is about as good as ever and I think I feel younger than I used to. I did not think I would live to be 40.

I am happy much of the time but the deep undercurrent of melancholy, the sadness of the world sufferings comes up into consciousness and make me sad. I want to find a religion that will reconcile me to these things. I think sometimes I will find it.

December 11, 1903
I can't seem to quench my thirst to solve the greatest problems. Will read something that will give a new view and make me happy. Gradually its strongest effect slips away. Not entirely, however, for I seem to hold a residue of everything strong. Only at times everything seems to be swept away and for a moment there seems, after all, no excuse for thinking we are better than other animals. God and immortality seem sweet dreams without any objective reality. But on the whole I believe my faith is growing.

January 21, 1904
I had a talk with Mr. Hatcher today and if our plans work, if I get a grant from the Carnegie Institute, I will be happy. I will then get a decent salary $1,500. That may mean much to Pearl and I.

Sunday, January 24, 1904

It seems far away to a comfortable home now. A day or two ago I thought I would soon be able to have Pearl come any time she wished. Now unless something fortunate happens it may not be for a year or more. Mr. Hatcher thought we could get a grant from the Carnegie Institution. Now I may not be able to get out West. May not be able to see my sweetheart for another long year. I stopped over in Homewood at the Presbyterian Church. Stopped after church at a real estate office. One of the men took me and showed me one of the new homes. It was nicely finished. Would like such a one, $4500; $500 down $35 per month.

January 27, 1904

Cold snow again. Our January thaw caught a cold.

My greatest desire now is to get a home. . . . We have been talking over the home proposition. I don't believe it pays to rent if one can buy a home. If Pearl and I marry I do not think either of us are extravagant and I believe we could get along nicely. My insurance is $15 per month. I could get a good home by paying $35 per month. We would not have to pay rent and I believe we could live on $50 per month. . . . Then, of course, we would have to have clothes. If I could save a few hundred dollars to get started and get $100 per month I believe we might venture.

February 28, 1904

Since I last wrote I have transacted some important business, the results may be far reaching but cannot be seen yet.

Mr. and Mrs. Peterson and I went out to Homewood Monday forenoon, Washington's birthday, to look at houses. I was as much impressed with the houses as ever. I told Mr. Hatcher about it and he said to look in to the matter carefully and if I wanted to buy he would let me have the money. I went out again Thurs. morning. Went to Mr. Olson. He had one that he would sell for $200 down. Did not exactly like the situation, though was tempted to buy. I then walked to Hermitage Street and saw an office in a dwelling. Made up my mind I would go in and inquire anyway. I inquired the terms and he said they were $200 down and $30 per month. "They are not like this one," I said. "Yes," he said. I went through the house and liked it very much. I talked with him for some time and told him I would return in the morning if I could. I went to Mr. Hatcher and arranged to borrow $150, $25 now and $125 in April. I went up there in the morning and talked the matter over and looked at several houses. Finally my choice lay between two corner houses. I finally chose the one highest up, next to the hill and the cheaper one of the two, and am perfectly satisfied with the choice. It will cost $5200. In doing this I felt I would meet Pearl's approval. She, like me, wants a home above all things. I know almost that she will fall in love with the place The only question is paying for it. This will be hard, I know, but I am willing to sacrifice all for a home.

Tuesday, April 19, 1904
It has been a month and a half since I have written in this book.

Have moved into my house. Am not established but camped. Did not get some money I expected on the first and it has caused me a lot of unpleasantness and trouble. Hope to get established soon. Have paid $200 on the house. I rent all except the attic rooms for $30 per month. I only have to pay that much on the house but expect to pay more.

Wednesday morning, May 25, 1904
Received a long letter form Pearl yesterday. . . . I believe she has good sense enough to forget the past and if I treat her right, as I intend to do, I believe we will get along nicely. My old love remains and it seems she is the source of every inspiration. The one thing, more than all others, that makes life worth living.

I am very busy now preparing notes for my next publication. Am trying by the present to get lights on the past. Slowly it seems to come. Yet of the White River times I have not been able to form a definite picture and do not know whether the climate was very dry or moist.

June 5, 1904
I did something yesterday that seemed rather foolish or hazardous. It may turn out to be nice or foolish. Time will tell. I did not intend to tell anyone but I engaged a lot. Paid $5.00 on it, am to pay the other $5 tomorrow, the fellow advancing the money. I was tempted, it was such a place as I long have wished. It was beautifully located in a grove of large oaks. Beautiful woods slope down the ravine below. It is almost ideal. Three big oak trees on it. The lots have been selling like hot cakes and most of them on the hillside. It would be such a pretty place to live and, as I said, I was tempted. I shall have to practice the most strict economy.

June 8, 1904
I am happy most of the time now. I have to economize with extreme care since I made my last venture. . . . For example I have 7 or 8 dollars to last me 2 weeks and 3 of that I owe if the man comes after it. It was for plumbing. . . . Well I have 2 meal tickets, $1.55 at Kings and $2.05 at the Cozy. So I may make it. This is not reckoning the $4.10 for the first weekly payment on my lot.

Sunday morning , July 10, 1904
Since writing here last Mr. Hatcher has passed away. It was not wholly unexpected yet it came as a shock. Such things make us think over again these problems of human destiny. It is hard when a man has toiled all his life with unflagging energy and enthusiasm to be cut off when he had gotten where he could make his efforts count, yet the loss is not his. It is the loss to his family and to science.

Mr. Hatcher died of typhoid fever the 3rd of July a week ago yesterday. A week ago this afternoon was one of the happiest half days I had spent for a long time, not knowing that then Mr. Hatcher was dead. It was so beautiful where I went.

He was buried the 6th in Homewood Cemetery. Professor Scott was here from Princeton. . . . Professor Scott, Dr. Ortman, Mr. Gilmore and myself were pall bearers.

The death of Mr. Hatcher was a great loss to my father. Not only had he persuaded my father to join the staff of the Carnegie Museum, but he had also been an able teacher, helping greatly in his fossil collecting. Mr. Hatcher had distinguished himself in the field of paleontology. He, together with Mr. O. A. Peterson and Mr. J. L. Wortman, had uncovered portions of two dinosaur skeletons (Diplodocus) during 1899 and 1900 at Sheep Creek, Wyoming. In 1901 Mr. Hatcher, by combining these two skeletons and a skull discovered earlier by Professor Marsh together with parts from the American Museum, had published the first sketch of a complete skeleton of diplodocus. The restored skeleton was mounted in the Carnegie Museum and was named Diplodocus Carnegie. Not long afterward a plaster cast of this skeleton was made under the direction of Mr. Hatcher and Dr. Holland, director of the Carnegie Museum, and in 1905 was presented to the British Museum by Andrew Carnegie. Later plaster casts of the same skeleton went to several countries in the world. One of these restorations now stands inside the Utah Field House of Natural History at Vernal, Utah.

With the death of Mr. Hatcher, whom my father considered the best fossil collector he ever knew, things were not to be the same for him at the Carnegie Museum.

July 29, 1904

Foggy this morning, a nice comfortable day. Started to walk to the museum. I read Darwin as I walked along (Voyage Around the World). Turned off 5th Ave. at Beachwood Boulevard and went up a street that branches off it. Got into a place where there were not good sidewalks. Thought I turned to the right, struck a street and turned again and found it was Linde, so I struck Pecan Ave. two or three streets farther back than where I left 5th. Came back where I had been. Got on a car thinking all the time it came down 5th but it went down Pecan beyond East Liberty so got off and struck for Center. Got lost but found myself, got to Center, thought I would take a car but it did not stop so walked over to Ellsworth and took a car getting to the museum about 20 min. after 8 o'clock. That is one of my greatest troubles now, absent-mindedness about little things.

Walked home this evening, stopped at one place on a hillside in an old orchard to write. Nature never seemed so pleasing to me as it does lately.

July 31, 1904

I have gotten along splendidly with my writing on my new paper. I have a chance to work now and am going to work. If I can do the work I have to do I can get where they cannot hold me here unless they pay me a decent salary. Dr. Holland treated me curiously tonight and it was a surprise to me but I probably deserve it. I have been too quiet and retiring. Such people are always looked down on.

August 22, 1904

Mr. Raymond and bride returned from New York and Mr. Prentice came back from his trip to Lawrence, Kansas and the Worlds Fair.

When I saw Mr. Raymond's wife today at the museum with him it made me long to have my girl here. I felt almost indignant and thought I was being deprived of my rights.

September 13, 1904 Monday

I have been doing writing lately. I have begun to take a story writing course with the West Press Association and my time is well occupied.

September 15, 1904

Life is ever changing ever different. I spend my spare time reading and writing. My health has not been the best but am better now and if I am careful I can keep well. This I must do, not so much for myself as for others. My old father, who cannot stay much longer, my poor suffering sister Nettie and for she who has promised to be mine and left all else for me.

Yesterday I sent a new lesson and a repeated one to West Press Association ... Did not feel well and it was far from what it ought to be, yet I am angry at the criticism. It is the spirit of it, and in spite of myself I despise him more and more. I will probably, about the next time, criticize the critic.

December 16, 1904

I am getting along as well as could be expected. I have paid more than the monthly payment on my house and have my insurance paid.

On the lot I bought in Westview I have paid $115. I am to give $650. Mr. Weis told me tonight that it is worth $750. I think I can get $800 in the spring.

February 20, 1905

I go out so much evenings and have so much to do in my story writing course that I do not do any writing in my journal lately. I attend meetings at the Unitarian Church Sundays and Bible class Tuesday evening and we have another meeting of a half a dozen people every Sat. evening. Last Saturday it was at Mrs. Stoffer's, then there are the meetings of the Academy of Science.

A short time ago I received a check from Mines & Minerals to pay for my article in that magazine, "Source of Gold in Alder Gulch." It was for $21.50, the first money I ever received for writing.

June 11, 1905

    The evening before was Bible study class. I gave a talk on the miracles. It was followed by a long and lively discussion.

    Yesterday afternoon was an event in my life. Mrs. Pringle and I went down to Vincent and Scott's and picked out the furniture for the rooms below. People said I had no idea how much it would cost to furnish a house so I was prepared, or partly prepared, for the worst. It was both Pearl's and my idea to get nice furniture when we did get any and I picked out nothing but good things. I think it will be the cheapest in the end. . . . I have not bought but have only selected. I probably will not have enough to pay for them unless I sell my lot. This I will probably do if I can and I guess I can.

    I am anxious to see the things in place and my Pearl in her place. I think they will please her. It will cost about $400 to furnish the rooms but I will not have to pay it all at once.

    I do not seem to accomplish much in my literary work now. There has been so much else to do and I do not have the time. If I go out West in July my time here is short.

With the entry of June 11th the regular diary for 1905 ends abruptly. (GED).

⌘ ⌘ ⌘

CARNEGIE MUSEUM:
Founder's Day 1907

For some time it was my opinion that my father's diaries for the years 1906, 1907 and 1908 were lost, but a book was later found entitled, "Diary of Earl and Pearl Douglass" in which an explanation for this is given. After his marriage he kept no regular diary of his own until 1909. From letters, field notes, and other material the course of events can be pretty well traced, however. (See Appendix #1; 1908 journal from Carnegie Museum)

After the house was furnished and put in readiness for his return in the fall of 1905 he and Mr. Raymond, the young invertebrate paleontologist mentioned before, left Pittsburgh together on a summer's field trip. The date of their departure is not mentioned but on August 4, 1905 they arrived by train at the town of Medora, North Dakota. This was the country known as the Little Badlands and was close to what is now the Theodore Roosevelt Memorial National Park. Their plans were to go to what was called the T. H. Timber Ranch, which was located some 40 miles to the south up the Little Missouri River, but the roads were in such terrible condition that the mail man, whom they assumed would take them there, would not take more than one passenger in his small rig. It was decided that Mr. Raymond would go to Glendive, Montana where he wished to do some collecting, and my father would make the trip alone. It was a hard trip and took a couple of days. He rode with the mail man to the T. H. Ranch and from there rode a borrowed horse to the main ranch where he had quite an enjoyable time. The people were very hospitable. He described it as being a large ranch and a very nice place considering the fact that it was so isolated. Some interesting fossil material was collected in the area. He left Medora for

Glendive by train on August 19 where he joined Mr. Raymond. They proceeded together to Three Forks and later Whitehall, Montana. In September they moved from Whitehall to the Little Pipestone.

> September 10, 1905
>
> When we started from Whitehall to go to the White River outcrop on the Little Pipestone the morning was pleasant. That was the last stopping place before going to see the one I loved, the one that was to be my wife. I wanted to see Pearl yet I had been so used to staying away that I did not fret only I thought she might get tired of waiting if I stayed too long. I was happy, though, when on Tuesday we started for Alder.

> September 19, 1905
>
> In a cabin by Mr. Ptomey's at the foot of Old Baldy above Virginia City. We came here yesterday. We camped at Mrs. Goetschius'. Arrived here last Wed. (Ptomey's) Pearl did not want me to go the next day and consequently I did not want to. Mr. Raymond went to Spring Canyon. In the evening Mrs. Goetschius let me take her horse and buggy and we went up and got Mr. Raymond. He was very successful and was jubilant. He had found crinoids and some interesting trilobites.
>
> Today we went to the top of Old Baldy and beyond. We had a fine view from there . . . could see to the south Sheep Mountain with its Carboniferous strata lying nearly flat. This rugged line of huge peaks stretches far to the S. W. in the Snow Crest Range. To the S. E. is a region of high grass land and forests and parks. There is only one rugged prominent peak and that is Black Butte. That is of volcanic rock and is apparently the throat of an old volcano.

> T.F. Roberts Ranch, Sand Creek , North Dakota.
>
> It has been a long time since I have written any notes, not since the rush of getting married nearly 3 weeks ago. We have been on the move ever since and our time has been pretty fully occupied. From Alder we drove to Sheridan

Friday morning Oct. 20th. We went to the parsonage and Reverend Badeon married us. We then went to the depot and took the train for Deer Lodge. . . . Professor Provience met us there.

A few days were spent in the vicinity of Deer Lodge, Montana studying the geology of the lake beds and collecting fossils. Father and Mother then went to the above mentioned place in North Dakota where a number of fossils were taken out and shipped to the Carnegie Museum, after which they continued their journey to Pittsburgh.

The entire year of 1906 was apparently spent in Pittsburgh, no field trip being made during the summer month. The book entitled, Diary of Earl and Pearl Douglass was started January 1, 1907. The following are some of the entries from this diary.

January 1, 1907 (written by Pearl Douglass)
Quietly spent the day at home, Earl recovering from throat trouble.

January 6, 1907 (written by Pearl Douglass)
Earl began work in new room at Carnegie Museum. He is naming and describing fossils from Little Badlands.

January 31, 1907 (written by Pearl Douglass)
Earl and I attended "The Grand" in the evening. Saw Mrs. Langtry.
During the month we attended church every Sunday but the weather was so bad and disagreeable that we did not take a single walk.

April 2, 1907 (written by Earl Douglass)
I have not kept a diary since I was married. It has not been neglected because there was nothing of interest by any means, for our lives were never before more full and important to us.

Pearl and I decided some time ago that we would get a book and both write in it, for many days I cannot write and she can. Our hand writing is different enough to tell who is writing at a glance if we do not sign our names.

We were married on Oct. 20, 1905 and we have been together constantly since that time with the exception of three or four weeks.

April 1, 1907 (written by Pearl Douglass)
The weather was an "April fool". We had enjoyed a week of real nice weather and on the morning of April 1 we found ice in our kitchen.

April 2, 1907 (written by Pearl Douglass)
Went to the museum, met Earl and we both went down to the city. Took dinner then went to the P.O. where Earl asked for information concerning the Civil Service Examination. I hope he pushes the matter along for I feel sure if he does there is something in the future better for him than he now has.

We then walked across the Allegheny River and I went down to Boggs and Buhl's store to do some shopping. This is the first time I have ventured alone since I came to Pittsburgh (Dec. 20, 1905).

April 3, 1907 (written by Earl Douglass)
It is getting a little warmer every day. Had an indignation meeting at the museum, four of us. Have found definitely that one man's salary has been raised to $2000 while other men just as essential to the museum are left with a meager salary hardly enough to live on. It is considered nothing less than an outrage. The way things are run at the museum is certainly surprising to those who have not been initiated.

We are thinking some of selling our house if we have a good chance.

Have sent for information concerning Civil Service Examination as I am thinking of taking an exam. for a position as Assistant Geologist.

April 5, 1907 (written by Pearl Douglass)
Last night we listened to Mr. H. W. Du Bois lecture on Alaska, "The Land of our Midnight Sun." The lecture was exceedingly good. I never saw better lantern slides. One thing I wish to remember is that he told us the first gold found in the Klondike region was found by a squaw while washing out her frying pan.

April 5, 1907 (written by Earl Douglass)
I have been helping get some things in the cases today. We will undoubtedly be able to get our exhibit in shape if we get a couple more cases. We have a lot of unique specimens but I do not consider the hall at all fitted for exhibition and the doctor will not let those who are capable of doing so arrange things, and to my notion and that of others, his taste and judgment are extremely poor. It is too bad that such a man must occupy such a place.

I have become more interested in my work of late than for a long time before. Have been getting out bones. Never had them out for study and comparison before. I hope to write a paper on horses soon if I do not go west. I have the following ready for publication:
1. New species of merycoidonts. (one of the oreodonts.)
2. Merycochaerus and a new genus of merycoidodonts.
3. Extinct rhinoceroses.
4. A paper on glyptosaurus (an extinct ancestor of the Armadillos)

Several papers could be ready in a short time:
1. I could, in a short time, prepare an interesting paper on the Horse from the Titanothorium Beds to the Madison Valley Beds.

2. I am beginning an article for a bulletin for the American Museum of Natural History on some Merycoidodonts from Montana. Dr. Mathews has sent me the two skeletons from the Upper Boulder Valley. This is an interesting little lot of material and ought to make a good bulletin.
3. I have also material for a fine paper on Oligocene rodents.
4. I have also some fragmentary marsupials and some insectivores which are interesting.
5. If I do not go out my larger paper on the Fort Union Beds could be ready soon.
6. There is also some colodon (extinct animals thought to be related to the horses and rhinoceroses) material that is of interest.

I have well under way a quite exhaustive paper on the Mesozoic and Cenozoic formations of North Dakota and Montana, with special reference to the Fort Union.

I have also 200 typewritten pages or more on the Oreodontidae of Montana.

Then I have a lot written on the geology of Montana.

I have recently become much interested in the Eocene formations and have made a request to be permitted to go to the Big Horn Basin in Wyoming or the Uinta Basin in Utah. I am wild to go to one of these places and make a fine collection of Eocene mammals and study the geology of the country. I have been so interested in it that I have dreamed about it nights. I do not know whether I will be permitted to go or not. If I am not I think I can be reconciled for I have such a fine lot of work to do here. The way things are arranged now, I believe if I had good health I could finish, or nearly so, the papers I have outlined above in another year, especially if we can arrange so Pearl can help me. After Founder's day I can probably devote as much time as I can stand to literary work.

Founder's Day 1907, a historic event in the history of the Carnegie Museum and Pittsburgh, was devoted primarily to the dedication of the new museum building as it now stands. The event was internationally attended and its effect on Pittsburgh and my father and mother are recorded in the diary.

> Saturday, April 6, 1907 (written by Earl Douglass)
> The papers announce that Carnegie has announced the amount of his gift to the Institute and Technical School. The amount is stated as $6,000,000, $4,000,000 of which is for the Institute. It looks now as if we could tell pretty soon whether our promises have been all chaff or sincere. If I am not treated somewhere near right in the way of salary I shall begin to prepare to get out. I feel pretty sore about the salary business and I fear I shall until I am treated right. . . . If he will not give a decent salary after I have been preparing for so many years I hope he will at least print my papers. We are expected to work tomorrow.

> Sunday, April 7, 1907
> The "Doctor" told us to come to work today so we went, most of us. It occurred to me after I got there that I might get flat cases and put in the types which Dr. Holland wished to put on exhibition so I worked all day, the Coggeshells helping me. Peterson helped some so we got it all in. I fear other departments are not so fortunate. I suppose we will have to work nights until Thursday. I am glad the crazy rush is nearly over. I never saw Arthur Coggeshell so disgusted and discouraged as he has been of late. He is ambitious to go ahead and get things done and the doctor, while in an awful rush, will not let the man go ahead and do the work.
> Thoughts:
> The view we are now getting of the past by discovery of fossil animals and plants make the present world ever new to us giving to everything a wider interest and a greater

significance. Every little untouched spot of nature, every tree, every plant suggests new ideas and is a little incentive for the world of the imagination.

I suppose I have realized it to some extent for a long time but it never seemed quite so clear to me before that the truest and highest pleasures of my life are those of the imagination. Of course it is a pleasure to me and a gratification to get warm when I am cold, to get well when I am sick and to have wants gratified, but after all, what makes all things real pleasure is this faculty of the human mind.

Saturday, April 13, 1907 (Written by Earl Douglass)
We have been extremely busy since I wrote last. About every moment has been occupied. The great eventful day is gone at last and the rush is over. I imagine it has been an event of considerable importance in the history of the country but there has been too much of it to dwell at length on here. It is recorded in the papers and probably will be published in a permanent form. During the last few days we were especially busy and worked part of the night.

The front part of the institute building is beautiful. The museum part might be much better or much worse. We are yet in suspense as to our fates. That is as to our salaries. I have been contemplating the U.S. Geological Survey but do not want to leave here. My heart is in the work and I want to spend at least another year here and round up that work. I do not feel that I will ever, now after waiting all these years, get the salary I ought to have.

A couple of quotations from Pittsburgh newspapers give some idea of the significance of Founder's Day at the Carnegie Museum in April, 1907. The following from the Gazette Times, Pittsburgh, April 7, 1907 is quoted in part:

"Blazing in refulgent beauty! Peerless, unique, stands the new Institute building. Noble in form, flawless in finish, the crystallized embodiment of a great idea, it awaits the moment when it will take its place as the chief jewel in Pittsburgh's diadem.

The elevation of its place of lofty splendor is to be accompanied by ceremonies that will send a thrill of interest from end to end of the civilized world, and will be attended by a throng of potentates in the domains of letters, arts and science, the like of which has seldom been drawn together before under any circumstances. They have journeyed thousands of miles, from distant countries, to pay honor, not so much to the man whose kingly wealth and broad humanitarianism made the Carnegie Institute possible, but to the nobility of the thought which has developed into such a majestic fulfillment, and to give Pittsburghers a due appreciation of the magnitude of the gift they have received, and the manner in which it is regarded by thinkers of the world.

They will gather within its halls next Thursday and inaugurate imposing services appropriate to the dedication of this princely gift, and three days will be given up to a dignified, impressive program in which some of the world's greatest students and men of achievement will participate. ...

The building as it stands today represents the fruition of a thought that was sown almost 26 years ago, and which, after a protracted period of uncertainty and doubt, finally shot up into a growth that has expended with a rapidity and vigor that has dazzled the whole world. It represents more than $20,000,000 given by one man to develop his kindly design upon his fellow citizens, and contains a store of priceless gems of art, science and literature that more than repays the expenditure."

The following is from The Index, Pittsburgh's Illustrated Weekly, April 13, 1907:

"With ceremonies of most imposing character the new Carnegie Institute was dedicated Thursday afternoon. A great company of distinguished men and women representing the chief nations of Europe and the American hemisphere took part in these ceremonies, which formed the climactic incident of a program of events beginning Thursday morning and not coming to an end until late this afternoon.

Pittsburgh has never, perhaps, lived through three more notable days, and in the history of education in America there are few to be compared with them, but it has seen, as well, what may be regarded as the world's first view of the Carnegie Technical Schools.

Thursday began with a reception of guests of honor by William N. Frew, President of the Institute's board of trustees in the Founder's Room.

In the afternoon came the most spectacular event of the dedication, the academic parade from the Hotel Schenley to the Institute. Between lines of students from the technical schools the guests of the day passed.

At 2 o'clock came the dedication proper, Mr. Frew presiding, greetings from President Roosevelt and from England, Germany and France lending distinction to the exercises. Mr. Carnegie made a brief address, as did Theodore Von Moeller, Minister of State of Germany and Baron d'Estournelles de Canstant, of the French Senate."

April 20, 1907 (Written by Pearl Douglass)

I have not written any since April 5, have been so very busy that it seemed I could not get to it. I put in the second week of April sewing. Sewed steadily during the day and in the evenings when Earl was away. He worked several evenings just before founder's exercises. I did not mind it much as I was so busy. I feel I have so many things to be thankful for and this one of them, that my dear companion is home nearly every night, I mean when we are not out together. I could never be contented if he spent his evenings out.

Founder's exercises began April 11, lasting three days. They were very interesting to me as I had never seen anything so elaborate before. I enjoyed seeing and hearing the foreign visitors. I had the privilege of hearing some of the important men of the day. Robert Ball seemed to create a sense of humor and W. T. Stead brought his independence to America with him. I enjoyed his talk and think his views were fine. So many are bitter against him but I believe they misinterpret him for he pointed out their corruption so strongly that they are compelled to take refuge.

I thoroughly enjoyed hearing Mr. Carnegie talk. He is a little man with white hair and a very strong blue eye. I believe he has a heart of sympathy for the people in general but I wonder if he feels that they realize as much from his libraries etc. as he intended they should. Sometimes I almost wish the dedication had never come. Sometimes it seems more pleasant to live in hopes of something better coming than to witness the realities of life. It seems that times will be more serious for us than they have been and they surely have been trying enough, but I believe there is a brighter day ahead for us but how far away and whether we will have patience to wait for it is another question. I feel so much for Earl. He is so patient and earnest in his work and receives so little pay for it. But after all I realize we have much to be thankful for. We have our health and each others love and sympathy and I am very thankful there is no one else to suffer this scantiness and privation.

Sunday, April 21 (written by Pearl)

This was our first long trip to the country this spring. I thought I could never enjoy such freedom and such a sense of extreme pleasure again as I did my first year in Pittsburgh, but I believe it is going to be just as grand as last year. Perhaps more so because I am a little better acquainted with nature than I was last year. I enjoy traveling around here immensely. The walk to the Verona Car Line was delightful. The birds seemed so glad and everything was beginning to move. I enjoyed the ride to Verona but the ride in the launch across the river was so pleasant, then we took the Natrona Car, which was new and nice, and I enjoyed this beyond expression. I looked at Earl several times and tried to tell him what I felt but could not. Such things make life more worth living. We found our first spring flowers. I drew my first picture from nature and painted it in water color.

May 31, 1907 (written by Earl Douglass)
We are somewhat anxious just now. I made application for a trip to the Big Horn Basin or the Uinta Basin. I have not heard the decision yet. I want to go to Utah. The U.S. Geological Survey is to send out a lot of parties to survey the coal lands. Dr. Knowlton and Dr. Peale are going to investigate the relations between the Fort Union and the Laramie. I wish I could go to investigate fossil mammals. It is what I have wished for years. Knowlton says they are not to have a vertebrate paleontologist. That is indeed unfortunate.

My father's request to go west was not granted and no field trip was made during the summer of 1907. Another year had to pass before his dream could be realized.

December 18, 1907 (written by Earl Douglass)
We have been getting along as well as could be expected. We are awaiting an event that will surely make a change in our lives. I don't know how it will affect me. I wait for the time to come. I have learned to live pretty much in the present. Pearl and I have spent two unusually happy years together. It has been a continual financial struggle but we have been patient.

January 8, 1908 (written by Earl Douglass)
Have got hard up and had to mortgage my lot. Cost about $52 and I only got $350. I don't know what we would have done if we couldn't have gotten it.

March 8 1908 (written my Earl Douglass)
When I wrote last we had been expecting the baby for some time but no real show of his coming. The Dr. got the

date too early and we came to think she was right. Pearl suffered a great deal of discomfort and it was a very trying time and we were anxious to have it over. She had some pains but we learned afterward it was not the real thing. On the night of January 30, we found by rechecking the thing it was not due until the 29th. She got up at midnight and began to feel pains coming very frequently. The nurse, Miss Margaret Fleming, had come two or three days before. After a while we decide to call her. The pains increased and for a time were only a minute apart. Miss Fleming telephoned the doctor. Miss Marshall was ill herself so she sent for Mrs. Hall. She got here about 4 o'clock. It seemed hard to see Pearl suffer so but she was doing the work finely. I went to the drug store and got some chloroform and the last two hours or so they gave her that so now she does not realize she suffered so much. The boy was born a little after 8 o'clock. We expected a small boy and were surprised at his size. He weighed 10 pounds and was a stout, healthy, well formed fellow. The doctor said, "Its a great big boy". Pearl said, "Oh my, I'm glad of that." She was determined to have a boy and would not talk girl. . . . From that time on it kept the nurse busy and me too when I was home and that was a good deal of the time. . . . Pearl took the grip in about a week. The Dr. feared at first it was the typhoid fever but her temperature came down the next morning. I think I was never so glad of anything in my life than I was when I found she did not have to have the fever. About that time Miss Margaret Fleming went away to attend to another case and Miss Laura Fleming came. Both girls are jewels. It kept them busy night and day but they were faithful as if it had been their own people. Pearl improved right along. . . . Mrs. Ortman and Mrs. O'Leary came and dressed the baby two or three times.

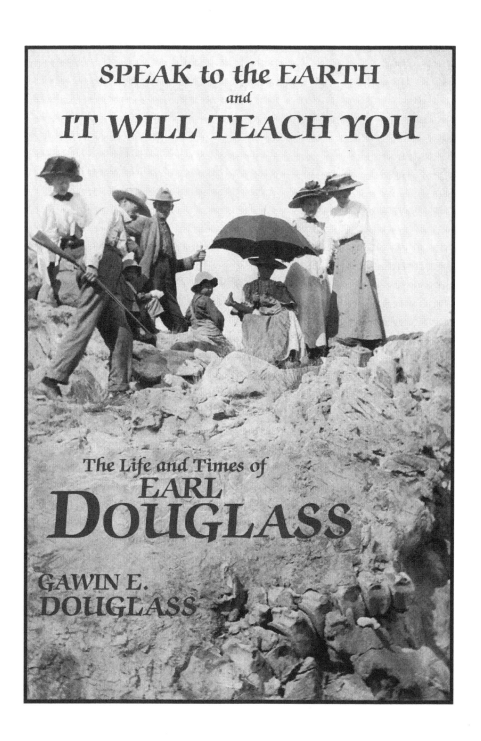

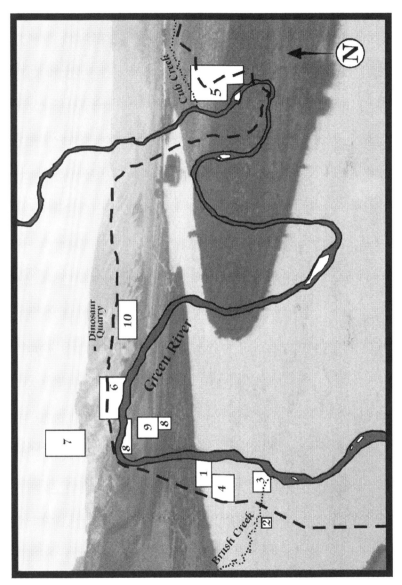

Figure *. Map of the Carnegie Quarry area, circa 1908 - 1920. Landowners mentioned in this book include: 1) William Neal, 2) Joseph Ainge, 3) Isaac Burton, 4) John Murray, 5) Edwin Lewis, 6 & 7) Earl Douglass, 8) Charles Neal, 9) R. C. Thorne, and 10) William Schaefermeyer.

1. The Douglass family about 1890 - Abigail, Nettie, Ida, Fernando and Earl.

2. Portrait of Earl Douglass in St. Louis - 1900.

3. Three generations of Douglass males: Fernando, Gawin, and Earl.

4. Brookings Institute - "Truth or Bust" group

5. Earl reading under the Lone Tree by the Green River.

6. Gawin playing catch with Joe Ainge.

7. Winter shot with Pearl on the skid driving horses and Golden York standing by.

8. Earl Douglass writing data sheets and letters for more money? - 1908

9. Jurassic beds with buggy in the foreground - 1908

10. 1908 - fibula from a dinosaur - fragments pieced back together.

11. Dr. Holland and Frank Goetschius (Pearl's brother) in the Uinta Formation in 1908.

12. Discovery vertebrae with Douglass and eleven visitors - it was popular from the very beginning.

13. Original August 17, 1909 discovery-eight tail vertebrae from a "Brontosaur."

14. View eastward from the Green River toward Split Mountain.

15. Using feathers and wedges to split the rock on the quarry face.

16. Caudal vertebrae with pelvis and pubic peduncle in quarry face.

17. Lowering of sauropod femur - three men with a block and tackle on the quarry face.

18. Pelvis with femur and caudals in place.

19. Lunch break with watermelon - pick a soft seat, any rock will do.

20. Earl Douglass and O. A. Peterson about 1920, next to the dinosaur wall.

21. Pearl, Gawin, Earl (mapping) as well as men working on the quarry face (1912).

22. Dynamite blast with numerous debris blocks - looking toward Blue Mountain.

23. Multi-tasking on the quarry face.

24. Quarry face with railroad tracks

25. Block and trackle with tripod

26. Two men pushing a cart up the tracks at the quarry (the only railroad in the Vernal area - ever).

27. Rail tracks covered with plaster jackets ready for transport to the Uintah rail head.

28. Prospecting the quarry face - Earl rappeling down the cliff.

29. Three men lowering plastered bones with a block and tackle.

30. Three men on the track, rolling the crate on a piece of pipe.

31. Partially plastered pelvis of a large sauropod.

32. Fencing used to stabilize the plaster on the pelvis of a large sauropod.

33. Lower the wagons, don't raise the bones.

34. Photographer capturing a camp scene at the base of the cliff.

35. Three men hazarding their lives to find dinosaur bones.

36. Team of four horses straining to pull a plastered block and two crates of bones.

37. Wagon train at "Alhandra" ferry crossing - south side of the Green River.

38. Team and wagon full of bones ready to disembark on the river crossing.

39. Bones under tarp on the Uintah Railroad going out of Dragon, Utah to Mack, Colorado.

40. Freight wagon with bones unloading to the flatbed on the railroad car.

41. Six teams and wagons with bones headed toward the ferry in winter.

42. Riding the skid loaded with a block of dinosaur bone.

43. Bill and Joe (the mules) plowing the base of the quarry. Notice the grid marks used for mapping.

44. Three wagons and a Model-T Ford loaded with bones by the Douglass cabin.

45. Prospecting on *Dolichorhinus* hill, worker lowered by rope.

46. *Dolichorhinus* quarry - 5 large mammals from the Uinta Formation.

47. Pearl and Gawin in the garden in Orchid Draw.

48. A view of Camp Gulch taken from the hill south of the quarry site.

49. Orchid Draw - Earl Douglass, Jay Kay, Clarence Neilson, Joe Ainge, O. A. Peterson, and young Gawin.

50. The Lone Tree (or Powell Tree) on Green River with the Kay cabin

51. Pearl pulling Gawin on the ice on a home-made sled with runners.

52. Pearl reading to an older Gawin in the shade.

53. Snow encroaching on the tent - the early years.

54. Earl reading inside their log home at the quarry - a little civilization in the wilderness.

55 Earl Douglass all dressed up, standing and contemplating - saying good-bye?.

THE WEST BECKONS AGAIN:
The Land of the Telmathere

My father's dream of going to Utah to collect fossil mammals now became a reality as his request was granted in the spring of 1908. To him the Carnegie Museum and Pittsburgh was a vast storehouse of recorded knowledge and springtime there was very charming, yet the hot barren wastelands of the West, where many secrets of the past were waiting to be unveiled, were much more attractive. In his poem, The Land of the Telmathere, he touched on the great fascination it held for him.

<u>The Land of the Telmathere</u>
    By Earl Douglass
The spring has come and the air is sweet
'Twixt the winter's cold and the summer's heat,
And the garnered wealth of the past is here,
But I go to the land of the Telmathere.

In the land where the Telmathere once throve
Are sage and cactus and yucca and sand,
And the bone hunter camps in no shadowy grove
And fierce is the sun in that desolate land.

Here, sweet airs flow over meadows free,
And the sounds of nature are pleasant to hear;
But the wonderful past is calling me
And I go to the land of the Telmathere.

Though the powers of the air for ages have played
In the winds that roar and the storms that beat,
And fanciful forms of the rocks they've made,
And faces that grin in the blinding heat;

For the skulls and the bones that here we find
Bring the balmy days of the past to mind,
Strange beasts and forests and streams appear,
And we live in the days of the Telmathere.

Note: Telmathere "marsh beast" - A beast that lived long ages ago in the Rocky Mountain region and whose bones have been found in the marsh and river deposits — which are now badlands — in Utah and Wyoming.

Preparations for the trip to Utah were made early in the spring of 1908. Mother and I were to accompany my father to Medford, Minnesota where a few days would be spent with his family before his departure for Utah. Mother and I, after remaining in Medford for a time, were to go to Montana where the remainder of the summer would be spent with Grandmother Goetschius. The latter did not transpire, however, as is explained in the following diary entries.

> April 11, 1908 (written by Pearl Douglass)
> This day we packed our household goods and left the house clear so it could be rented. Gawin has been extremely good for a 10 weeks old baby. At 8 o'clock we left Mrs. O'leary's house, where we had supper, for the Union Depot. Here we took a train for Chicago starting for Medford. We took the Pennsylvania Railroad. This was Gawin's first ride on the train and we were anticipating trouble but he slept all night fine and did not awake until five next morning.

> Saturday, August 8 - 1908 Medford Minnesota (by Pearl Douglass)
> I have resolved that every night when I go to my room I am going to at least record the events of the day.

After the usual work of the day Nettie played and sang some at the organ. After dinner was over Father, Nettie, Gawin and I drove to Alfred Battin's to attend Sabbath School. I have felt so lonely and lonesome for Earl of late. Time is passing quickly but it is long at times. This separation is hard, we both feel it. Gawin is one of the best babies.

August 9, 1908 (by Pearl Douglass)
After dinner Nettie, Father, Gawin and myself drove up to Mr. Wolford's. Got a few blackberries and strawberries. On our way home we gathered leaves from all the native trees we could find. Am getting these for Earl to compare with his fossil leaves. I was preparing to start for Montana the 12th but word came from Mother this morning that they had the mumps at Uncle Will's, so I postponed the trip on Gawin's account. It seems I am destined not to see mother this summer.

I find very few notes in the files regarding my father's activities during the summer of 1908. This was, however, as we look at writings of a later date, a very productive summer for fossil mammals from the Uinta formation (Upper Eocene) in the Uinta Basin in northeastern Utah. A manuscript written by my father years later entitled "The Dinosaur National Monument" reveals more regarding the summer's expedition than anything else found to date. Under a subtitle, "Some of the Carnegie Museum's Explorations in the West" he wrote the following:

"After spending many years in collecting fossil mammals – camels, three-toed horses, rhinoceroses, mastodons, and many other – in the upper Tertiary deposits of western Montana I made a request of Dr. W. J. Holland, the director of the Carnegie Museum, that I be sent to Utah to make a collection of fossil mammals from the Uinta formation there. In the spring of 1908 I started for the field.

I procured a team and outfit in Grand Junction, Colorado. My brother-in-law accompanied me. From Mack, a station on the Denver and Rio Grande Railway and the southern terminus of the narrow-gage Uinta Railway, we drove to Achee (Atchee), which is at the foot of the great escarpment, which seemed to shut out the long-dreamed-of hunting ground of the fossil collector. While looking at this great rocky curtain and wondering how we were going to get over it we saw two dark lines of smoke arising from just below the top of the rim, then a dark thread, and the truth suddenly struck me and I exclaimed, "Great Heavens! is that where the train comes from?" "Heaven" was the right word, for the train, a double header, was just starting down from the sky up in the region of Baxter Pass.

We were advised to ship our wagon over the pass on a flat car, and it was good advice as it was enough for the horses to climb the steep slope without the wagon and load. It was only about three miles to the top of the pass by the trail, while the railroad (said to be the steepest in the world without a cograil) twisted and "squirmed around" as it climbed a distance of seven miles to the top of the pass. It seemed like a great serpent that had tried to climb over and had given up.

At Dragon, which was then the terminus of this railroad – fifty-five miles in length – there was a dense, black smoke issuing in clouds from a large, straight, perpendicular fissure in the sandstone. The filing of the fissure was a black, jetty material and resembled coal but was much more bright and shiny. It could be plainly seen that it would burn. The fine dust will explode something like gunpowder and this was the cause of the fire in the vein. It is the dried asphaltic residue of petroleum and is called gilsonite.

From Dragon we followed the Uinta Railway stage line to White River, which flows westward through the Uinta Basin and enters the Green River, which is really the upper portion of the Colorado River. For 25 miles or more the White River has cut picturesque canyons through the Green River shales and overlying Uinta deposits. The shales are nearly 1000 feet in thickness.

North of this canyon is a region of desert prairies with outlying buttes and mesas, and beyond this, to the northward, is a wilderness of badlands colored yellow, brown, green, grey and red. It was in this region, we did not know just where, that we hoped to find the remains of extinct mammals found nowhere else. We camped in a stone cabin by a gilsonite vein, and began to search the sandstone ledges in the surrounding buttes, mesas and bare badland areas, and not without success. We remained in this camp for five months and obtained two four-horse loads of fossils in the rocks.

During the summer Dr. Holland came to visit us in our camp and to look over the fossil grounds. He was much interested in the big dinosaurs and he went with me to some bluffs near the Green River where Mr. Burton, who was superintending the construction of a gold dredge by the river, had shown me some large bones. Dr. Holland hardly thought that the prospect warranted extensive excavation. On our return we crossed the river on a ferry and followed a wagon trail between the Green River on the south and the bare, rugged, fantastically carved Split Mountain uplift. Instead of seeking an easier route through softer rocks on the west side of this uplift, the Green River has cut a deep canyon through the older, harder rocks in the heart of the mountain. We little thought as we rode along admiring the unique scenery and remarking that it was different from anything that we had seen, we were passing within a half mile of just the thing the Carnegie Museum and other museums had sought for many years, a greater burying ground of prehistoric monsters than had been found – one which contained complete skeletons and perfect skulls of great dinosaurs."

*** Editor's note – Gawin Douglass did not have access to the journal kept that summer, so data was not included in the original manuscript of this book. A transcription of the 1908 Field Journal (done by Elizabeth Hill of the Carnegie Museum is included in an Appendix and greatly enlightens us regarding the activities that first summer in the Uinta Basin, SAB

Dr. Holland in his Carnegie Museum publication, "Earl Douglass, a Sketch in Appreciation of His Life Work" Issued June 30, 1931, mentioned his visit with my father in 1908 and related his memory of their search to see if there were indications of valuable dinosaur material in the area.

> "As we slowly made our way through stunted groves of pine we realized that we were upon Jurassic beds. We tethered our mules in the forest. Douglass went to the right, I to the left, scrambling up and down through the gullies in search of Jurassic fossils, with the understanding that if he found anything he was to discharge the shotgun which he carried, and if I found anything, I would fire the rifle which I carried. His shotgun was presently heard and after a somewhat toilsome walk in the direction of the sound I heard him shout. I came up to him standing beside a weathered-out femur of a Didplodocus lying at the bottom of a very narrow ravine into which it was difficult to descend. Whence this perfectly preserved bone had fallen, from what stratum of the many above us it had been washed, we failed to ascertain. But there it was, as sound and perfect as if it had been worked out from the matrix in a laboratory. It was too heavy for us to shoulder and carry away, and possibly even too heavy for the light wheeled vehicle in which we were traveling. So we left it there, proof positive that in that general region search for dinosaurian remains would probably be successful."

The last entry in the diary of Earl and Peal Douglass tells of the return to Pittsburgh, after the long summer in the Uinta Basin of Utah.

> November 13, 1908, Pittsburgh (by Earl Douglass)
> We, Pearl, Gawin and I have recently returned (Nov. 9). As is usual, some of the most busy and interesting portions of

our lives are not recorded in writing. Life is ever changing and the last 7 months of my life have been different from any other period. We have been married over 3 years now and of course married life is different from single life; happier and fuller though I had been very happy at times before.

I was separated from Pearl and my own little boy for about 6 months. He was only 3 months old when we parted. I do not care very much for a little baby. I thought when he came he was much better looking than other babies I had seen but he seemed, much of the time, a bunch of noise and ill temper. Yet I was interested in studying for his future.

Thus my father's much-anticipated field trip to Utah ended and the Carnegie Museum was further enriched with an excellent collection of fossil mammals that would provide interesting laboratory work for some time. This trip gave him only a glimpse, however, into a new and fascinating land and he would soon be making plans for an expedition into the same country the following summer.

This new land was also to cause the end of his scientific publications, which had been numerous since becoming connected with the Carnegie Museum. Dr. Holland in his Carnegie Museum Publication, "Earl Douglass, a Sketch in Appreciation of His Life and Work" listed the scientific papers he had published up to this time and commented on them: "A study of the foregoing papers reveals that he added seventeen genera and eighty-three species to the ever-growing list of fossil vertebrates."

⌘ ⌘ ⌘

27

EXPEDITION 1909:
Carnegie Wanted Dinosaurs

The fine collection of fossil mammals made by my father the previous year, together with the prospect of dinosaur material, made the Uinta Basin an attraction for further exploration to those directing the progress of the fast-growing Carnegie Museum in 1909. The museum was now approaching an equal rank with the largest museums of the country in the field of paleontology. During this period in the history of its study important discoveries were being made rapidly and competition between the larger museums was keen. The collection my father had brought with him to the Carnegie Museum had helped round out its fossil mammal collection considerably and the first complete skeleton of Diplodocus, restored by piecing together portions of several skeletons, had brought it much notoriety in the field of dinosaurs.

With the spectacular skeleton of Diplodocus standing in the large hall in the recently completed museum building, it now seemed fitting to those in charge, especially Carnegie, that more complete skeletons be added if possible. In those days complete skeletons of Jurassic dinosaurs were practically nonexistent and, in fact, it was only speculation that they would ever be found.

Although my father's primary interest in paleontology was not dinosaurs but mammals, he nevertheless had a keen interest in them. He once related that during his early collecting days in Montana while trudging up a hot, barren, canyon he had a vision of suddenly, upon rounding a bend in the canyon, beholding a complete skeleton of a dinosaur some thirty feet in length, weathered out in bold relief on the canyon wall. He said he never expected anything like that would really happen but the dream helped relieve his discomforts caused by heat and thirst.

In the early spring of 1909, while discussing with his fellow workers the probability of complete skeletons of Jurassic dinosaurs ever being found, he more or less boasted that if he were given the means and the time to search for them he could find them. At the time he little guessed he would soon have the opportunity to do this search and it would practically be forced on him.

His diary, which had not been kept for nearly three years except for a few entries in the Diary of Earl and Pearl Douglass, was again resumed in the spring of 1909. The following entries, which begin as he was about to depart for Utah, are taken from this book.

> June 4, 1909
>
> Got this book today at McClay's. Wanted a handy notebook to take out west. Do not always expect to cling closely to days and dates.
>
> There are many things I wish to remember and keep something like the sequence of events. Sometimes one event a day will serve as a string to hold many remembrances. I have made a mistake in not having this in the past.
>
> At museum - I realize I must regulate my work and make it more systematic. For the keenest, freshest thought, the morning is the time, when I first get to the museum, especially after a morning walk. I want to arise soon after 5 a.m., have breakfast a little after 6 and get started for the museum from 6:30 to 7:00. Then I can come, or walk to the museum and do a splendid half hour of work. What could a fellow not accomplish in that time. To do that I would have to arise early.
>
> I do not expect, however, to be here much longer this spring. We expect to go west as soon as we can get things straightened around. The most serious hindrance has been to get someone to go into the house and take care of it.
>
> Sunday, June 6, 1909
>
> Mr. & Mrs. McHugh were here today to see about the house. They liked it and we liked them. It is understood that

they will come here and stay while we are gone using what rooms and furniture they wish. When we return they will either move to the second floor or move out. I think they are just the people we want. They give $16 per month.

Wednesday, June 9, 1909

Am still at work getting things straightened up at the museum. Am closing up laboratory work. Have skull of Dolichorhinus turned over and am working on teeth. Was happy to find the teeth had not been destroyed.

Developed some films that are pretty good. I need to make a radical reform before I am a good photographer. I make so many blunders. I do hope something will turn up so I can get a good camera this summer.

Dr. Ortman and wife came out and took dinner with us and we had a pleasant evening.

Wednesday, June 23, 1909

Told Dr. Holland I was ready to go and wanted to start the evening of the next day. "All right," he said. He asked if anyone was going with me. I said no and that my wife was going to Minnesota and then to Montana. It was the first time I could get him to talk. He has seemed very peculiar about the expedition this summer.

Thursday, June 24, 1909

Was very busy all day getting ready to go west. Mrs. McHugh came the day before. She helped Pearl get ready. In the morning I went to Dr. Holland and handed him my manuscript, "Eocene Titanotheres." Gave him my estimate on cost of outfitting.

Team and harness	$400
Wagon	100
Wages for Asst., outfit, etc.	100
Transportation	75
	$675

He said it was too much for the team and gave me $650.

We got started in good time for the train but I left one grip on the street car when we transferred and a young fellow kindly brought it to me. I never was more grateful. Then we came away without checking our luggage.

Father, Mother, and I arrived in Medford, Minnesota June 26th. Father visited with his relatives and friends until July 9th at which time he departed for Utah.

Friday July 9, 1909

Was able to finish up pretty well what I wanted to do in Medford so I came away quite satisfied. Pearl, Nettie and Father with Gawin went up to the depot with me. Pearl stood the parting better than ever before.

Saturday, July 10, 1909 - at Alton, Iowa

Have been detained here on account of washouts ahead between here and Sioux City. Had a very heavy rain here last night.

We ran around N.W. Iowa and waited all day to get from Alton to Sioux City. Went a round-about way to get around washouts. Got acquainted with a Mr. C. C. Cobb, a merchant from New York, who had traveled much.

Sunday, July 11, 1909

Took the morning Denver and Rio Grand Train for Lincoln, Nebraska. There Mr. Cobb left me. Got a room at the Lincoln Hotel and had a good sound sleep. Took evening train for the west.

At Hastings, Nebraska I inquired of two trainmen how long the train would stop. I wanted to go to the drug store and get some medicine. I ordered brains in the diner. I noticed they were sour but thought they had put vinegar on them. I expected to have time and went to get some medicine as I had nothing in my grip. Before I got out the train started.

I tore after it nearly killing myself but too late. Stopped at a hotel and slept as good as if I had been on the train. My dreams were those of a mariner.

Monday, July 12, 1909

I took a morning train for Denver. I lost my berth but got a daylight ride so saw the country and located some Miocene deposits. May stop on my return. Had 1 1/2 hrs. in Denver.

Left on evening train for Rifle, Colorado. That road is about the bummest I know. They ought to pay $25 damages to every passenger who travels on it. ... The cars were crowded. There were lots of children and it was a marvel how patient they were.

Tuesday, July 13, 1909 – Arrived in Rifle around 9 o'clock.

Did not get my trunk to the hotel until afternoon. They did not have any prospecting picks here and had to send for some. Did not go out prospecting. Made some acquaintances.

Wednesday, July 14, 1909

In the morning took hammer and sack and walked down the track to get to some bluffs west of here. Went to the bluffs and thought they were probably upper Cretaceous. Did not find any fossils in those nearest the tracks. Went up a ravine two or three miles west of here toward the north and at a higher level found fragments of bones but could not tell what they were. A little father west I found some more and rejoiced to see that they were mammals.

I returned awfully thirsty and very hopeful. In a new country one seldom finds any such good indications so early.

Friday, July 16, 1909

Got on the Midland train at about 10 a.m. or after. Had a ticket to Morris. Conductor said train didn't stop at Morris, I would have to go to Rulison a mile or two beyond. I said, "I bought a ticket to Morris" and I was studying geology. "No sir, I'll put you off right here." "Do so if you think best." The train stopped at Morris. I searched the hills until nearly time for the train to return. Got the mandible of a rodent and part of the skull of a four-toed horse. The bluffs are steep and high and difficult to prospect.

Friday, July 17, 1909
   Got a saddle horse of Mr. Reese and started to prospect the country to the southwest. . . . Went over to Mamm Creek. Engaged a cabin on Mr. White's place.
   Had quite a visit with Mr. White. His father is a wealthy merchant in Boston. Mrs. White is homesick and wishes to go back East and wishes very much to sell the ranch. He wants to buy an old colonial mansion in New England. One that has the air of antiquity; does not like this country.

My father continued his collecting in the area south of Rifle, where he found interesting material but nothing he considered of great scientific importance. He returned to Rifle July 22nd and the next day found some fossils that he did consider of importance.

Thursday, July 22, 1909
   It was a little tiresome riding home as I had a load and did not go off a walk (the pace of his horse). Got to Rifle at about 1 p.m. Got a letter from Pearl saying that she was going to leave for Missoula, Montana the 23rd.
   I found it hot in Rifle as usual. My room is very warm.

Friday, July 23, 1909
   I got Nellie from Reese livery barn, the mare I have had since the 17th, and went north on the Meeker Road. Felt rather despondent most of the day as things do not look very good or encouraging.
   Went to Rifle Canyon after examining some soft dirt-like stuff. In the canyon rocks stand straight up. The coal beds

occur there. Found some men cutting an irrigation tunnel through the rocks. Above two other outfits were cutting other tunnels and out on the flat, men with scoops were cutting a big ditch. Examined some beds south of the canyon and found two pieces of jaws. I thought it looked like euprotogonia (one of the ancestral hoofed mammals, SAB) It was near the Upper Cretaceous and I thought it might be Fort Union. Got a little enthusiastic. As I started to return, around a ledge of cliffs, I found a coryphodon (one of the earlier known hoofed mammals, SAB) and a part of a carnivore. Got enthusiastic.

While traveling through Colorado during the summer of 1965 I decided to stop in Rifle and see if it might be possible to locate the spot mentioned above. I got up early in the morning, after spending the night in a motel, and without inquiring about the country started to look for it. I had never been in Rifle before but I thought I recognized the canyon from a distance. I found the vertical formations, the coal beds and the ditch, but could find no place where a ditch went through a tunnel. I was sure this must be the place but could not understand why there was no tunnel. I was about to continue my journey when I noticed another canyon further to the west and decided to investigate it before leaving. Upon entering the canyon I immediately saw a ditch. Although it had been abandoned for years it was plainly visible and a huge vertical sandstone strata that protruded out into the canyon had been tunneled through, no other way to build the ditch being feasible. I then had little trouble finding the spot were my father discovered the fossils.

Saturday, July 24, 1909
I got a span of cayuses and buggy and Mr. Grant and I went over east to the region where I was last evening. I found the broken skull of a carnivorous animal (creodont, a primitive flesh eater, thought to be the ancestor of the wolves and bears.)

> I think there would be work here for the rest of the season but I have about decided that it may be best to go and get the dinosaur bones near Vernal, Utah and look over some more of that territory. I may, however, decide to stay here and go there later.
>
> Monday, July 26, 1909
> In the morning getting ready to go west.
> The train was late, as usual on the Rio Grande, and was crowded. What I saw on the train will probably change my plans for the summer. The Wasatch stretches continuously from east of Rifle to west of Dubeck and there is no reason why it should not be fossiliforous as at Rifle.
> Where it ends this side of Dubeck and lies above the supposed Mesa Verde is the best looking of all, a veritable badland of bluffs. The prospect that it may be Fort Union is enough to startle a vertebrate paleontologist. Anyway it is probably old Wasatch.

I am sure my father's greatest interest in paleontology at this time, and perhaps for much of the remainder of his life, was the study and search for fossils of the early mammals that first came to light, in their true form, during the transition period between the Mesa Verde, the latest formation in the Mesozoic Era (the Age of Reptiles) and the Wasatch (Fort Union), the oldest formation in the Cenozoic Era (the Age of Mammals). Of course the fact that fossils of these early mammals were so rare at that time made the task of tracing them back to their beginning and speculating on the climate and topography of the country in which they lived, much more intriguing.

In coming from the east by train to Grand Junction, Colorado the previous year he passed thorough this country. It was probably from the train window that he had seen the rock exposures that had induced him to stop in Rifle and prospect for mammals before entering the Uinta Basin to continue his search for mammals and dinosaurs. Later entries in my father's diaries indicate that his great interest in fossil mammals was not entirely shared by those in charge of the Carnegie Museum.

Wednesday, July 28, 1909 - Written August 4

I think it was this day that I took the car to go to Myton. The car left between 1 and 2 p.m. They have only one passenger car and that is half used for baggage and mail. It was not a very elegant looking affair either. There were two women, sisters, going to join their husbands who had bought or taken up land near Myton. Then there was a tall young lady who I had seen at the hotel at Mack and impressed me more than most girls do now. She seemed quiet and refined.

It pretty nearly made me nervous going up Baxter Pass from the other side with a train. Once it stopped and it didn't seem as if they would ever be able to start.

A lively German and his wife are keeping the hotel in Dragon. They sat me down at a table and afterward sat the tall fair young lady at the same table. I did not think I had the right to speak to her but soon she introduced the conversation and we were friends afterwards. She has been going to school in Salt Lake City and is going home to teach in Vernal so we were on the stage together the next day. These things did not affect me except as a story writer who writes for magazines (ha! ha!) but if I had been young and single, well it might have been different. Then there would probably have been a previous fellow in the case and jealousy and wakeful nights and I know not what. I rode in the same stage to Bonanza. The latter place I found deserted and the greater portion of my things gone. All of my bedding; I will have to buy new.

I got in to Duchesne Thursday night after 9 o'clock I think it was. Some Chinamen keep an eating house there and I had a good supper. Found a bed in another house.

Friday I took the stage for Myton. Got there between 11 & 12 a.m. Went to the Calvert Hotel for dinner then got a horse at the livery stable and rode out to Mr. A. O. Smith's. Found him just the kind of a looking man I thought he wasn't. Mr. Smith had found a jaw of a large titanothere in the irrigation canal. The teeth were injured. It may have been diplacedon. I did not find anything valuable there.

We planned to take the team and buckboard and go south of the Duchesne River Sunday where Mr. Smith thought there would be good fossil hunting and collecting. We did not get started until late.

We had a little bay mare, a larger but poor sort of a horse, an old buckboard, a roll of bedding provisions etc.

As usual I started out with good hopes. Anyway I would see some country that was new to me. I had wondered when I left Ouray going out last year if anyone knew the desolate country to the S. W. beyond the Green and Duchesne Rivers and if there was any way of getting through it.

From Myton we struck to the southwestward. Passed one fairly good ranch run by the Colthorp Co. We then turned to the southwest on a road which leads to the Parriot Mine (Pariette, a gilsonite mine near the place where gilsonite was first discovered, SAB). After crossing a bench we came where we could look out over a peculiar badland country. It looked extremely favorable and I thought that surely we would get some good fossils there. We looked for some time but found nothing but fragments.

In the morning we headed eastward toward Ouray. We soon struck a ravine and went by a well where there was fairly good water except that there was a dead rodent in it. I examined all sorts of rock ledges of sandstone and reddish clay beds but as usual found nothing.

We went down the ravine for many miles. The sun was hot and the traveling was not good for the horses. I dug for water in the sand and would have gotten it if it had not been for rocks I could not get out.

In the afternoon we continued down the ravine stopping once below where two fences from cliff to cliff made a horse corral and digging in the sand for water below we found an abundance for the horses. Was too salty for us.

We drove down to where Willow Gulch enters the one down which we went then turned off and made our way toward the gilsonite mine which Mr. Smith and son had worked years before. We were anxious about water and had been speculating whether it had rained enough to fill the vein. I felt that it was pretty risky but where we stopped first, just as we were going to another place, I exclaimed "There's water." It was a little place that had been excavated for that purpose near a camp.

The mosquitoes are fierce in this arid land where it is many miles to water and I did not sleep much until after midnight. Toward morning (August 3rd) it got cold. Found no good fossils. In gravelly places at the bottom of sandstone ledges, where near Kennedy's Hole thought we would find bones, I found only impressions of vegetation.

We returned by another way after a while, getting out of the ravines and riding for miles over a bench strewn with dark pebbles.

Took supper by an old shaft, just before dusk, and then drove to Myton, camping or sleeping by the livery corral.

Written August 8 – In the morning (4th) I received several letters. One from Dr. Holland which made me change my plans, as he wished me to dig up dinosaur bones east of Vernal. He seemed somewhat disgusted, made me a little hot. Seemed to think I wasn't doing much. He had received the letter which I had sent to Stewart, which as the latter did not understand such things, was not very enthusiastic. I gave Mr. S. check and Calvert wouldn't cash it. Finally I telephoned Mr. Meagher (banker in Vernal whose acquaintance he had made the previous year, SAB) and he let a man in Colthorp's store talk with him and thus got the check cashed. Mr. Wall in Calvert's store felt very bad when he found out who I was.

Took stage at about 1:00 p.m. and stopped at Ft. Duchesne that night.

The next morning (5th) early, started for Chapeta Wells. Rode 18 miles and took breakfast at Ouray. It was about 1 p.m. when I arrived in Chapeta. Prospected around Chapeta

Wells in p.m. and found one valuable specimen. I think it is triplopus (an early ancestor of the rhinoceros.)

The next day (6th) rode on freight wagon to Bonanza, got things I had left there of my outfit, got them in freight wagons and took the stage for Vernal. My bedding and other things had been stolen.

The next day, August 7, spent most of the day purchasing supplies for my work. In the evening took dinner with Mr. & Mrs. Meagher. Spent a very pleasant evening for they are very pleasant and interesting people.

Sunday, August 8, 1909

I went out and took a walk in the country north of Vernal and took a few pictures. It was hot as blazes. It seems to me that it is always hot in this country, but I suppose it isn't in the winter, and when you get in a cool place near the water and in the shade the flies and mosquitoes and gnats are there to persecute you.

Well I walked a little way and stopped in the shade and lay down and had a nap, and the ants crawled all over me.

Stopped on my return to inquire where Mr. Daniel's ranch was and the young man said that was it. I had heard that he had a lot of bones. Young Mr. Daniels showed me remnants of dinosaurs; nothing good.

Monday, August 9, 1909

I expected to start for the dinosaur beds near Brush Creek and Green River Monday morning, or expected to until a lady at the hotel said Mr. Goodrich, of whom I had engaged a man and team, had been there and said that the young man who was to go with the team could not go. I was much disappointed. Old Mr. Goodrich came in and said he could get me down there and the young man could come later.

I spent the day getting things I had not purchased Saturday and finished preparations.

Tuesday August 10, 1909 - Written 13th

Mr. Goodrich came with team and wagon and we got loaded up and started for Mr. Neal's ranch where I expected to camp. I think it was after 1 P.M. when I got there.

We got lunch and then started for the dinosaur beds. Found that some one had taken the best of the bones that were exposed. The femur that was lying in position was about half gone.

Wednesday, August 11, 1909 – Written 13th

Went to dinosaur beds but things didn't look very favorable. Dug for a while and cleared up the ground some to get at the layer that contains the bones.

We returned about 1 o'clock so Mr. Goodrich would have time to go home. After lunch he started home and I went afoot to a gulch which comes out of the hills a little up the river from here. Found fossil shells in the so-called Dakota Sandstones then in the underlying gray shales found fish scales. In the underlying red and grey beds were petrified logs and bones of dinosaurs and still below shells of mollusks, belemnites. This will be a good place to get the relation of these dinosaur beds.

Thursday, August 12, 1909 – Written 13th

Went out prospecting again coming out of the gulch I went in the day before. Found dinosaur bones but nothing good. Saw broken remains of a little fellow. Once in awhile one can get a good limb bone here and I do not doubt that there are good specimens to be had but they don't appear to be very common.

Friday August 13, 1909

Went to the bone beds that Mr. Burton showed me when Mr. Goodrich and I began the other day and dug for some time. I dug above the bone layer and removed the shale for a little space. The layer contains fragments of bones but nothing especially promising. I am going to keep digging at it for a while anyway. Maybe I'll find some good bones.

In the mean time, until my man and team come, I will prospect other exposures. Another thing that I wish to do is get the approximate thickness of the beds.

It has been a month since I got into Rifle. Have prospected most of the time. I have, I think, at least learned where to go . If I don't find anything in one place I will go to another. I think I will get something good.

Saturday, August 14, 1909 - Written 17th

Mr. Goodrich and I drove down or up to the Red Wash to strike the bone beds there. I found where there were a lot of fragments of bones in a harder layer in the clay beds. Found also bones of a little dinosaur. I thought it looked as if there was a little dinosaur buried below. Struck the beds below also where there were lots of shells and belemnites.

Sunday, August 15, 1909 - Written 17th

Was hot in the forenoon, yes, until night, as it has been for days before. It seemed there was no rest. Mr. Beer, Mr. Goodrich and I went north of here to examine a ledge where the Jones had got out some bones. It was along the same bluffs where I had been digging at the animal Mr. Burton found but further north. There were fragments of bones in abundance but nothing that I thought would pay for getting out of the hard flinty rock. In fact nothing was perfect.

Got an Indian mortar with a pestle in it in the sand among the cedars just below the dinosaur beds.

Monday, August 16, 1909 - Written 17th

Went out and dug this side of the Red Wash where we found the layer of bones. Found that it extended further than I supposed but when I dug into it, it was disgusting. The bones were terribly broken up and it seemed as if they had been churned up. Found some little bones pretty good and uncovered a femur 5 ft. long that seemed rather slender.

Felt rather discouraged. Mr. Neal says, "Oh well , you'll find something." He seemed perfectly confident that there was something good and that I would find it if I didn't get discouraged. He is very much interested. His talk really encouraged me.

In spite of hardship and discouragement my father's vision of a compete skeletal form of a dinosaur persisted. Each day brought a renewed hope, a renewed faith that somewhere nearby his vision would materialize.

⌘ ⌘ ⌘

28

A DISCOVERY:
A Vision, a Mirage?

Thursday, August 17, 1909

Concluded that I would not dig away at fragments until I found out whether there was not something better so we drove to the gulch this side of the one we had last examined and I began searching the hard sandstones which are unusually thick and well exposed there. Examined both sides of the sandstone beds thoroughly and found lots of broken and shattered bones but as usual, nothing perfect.

At last, in the top of the ledge where the softer overlying beds form a divide, a kind of saddle, I saw eight of the tail bones of a brontosaurus in exact position. It was a beautiful sight. Part of the ledge had weathered away and several of the vertebra had weathered out and the beautifully preserved centra lay on the ground. It is by far the best looking dinosaur prospect I have ever found. The part exposed is worth preserving anyway.

Wednesday, August 18, 1909

Mr. Goodrich and I drove up to the brontosaurus, or as near as we could get to it, and began digging. I worked pretty hard. Was anxious to see more of it to know how it lies and how good the indications are. Found the ilium probably almost exactly in place. Dug down to what I thought was the femoral articular surface. Got a little lower and struck what I thought was the femur in exact position. But after digging around it a little I thought that it was perhaps an ischium.

Studied a little how I could make a snake road down to the ravine so we can get the wagon to the steeper slope.

It is natural for one to picture finding a whole skeleton, head and all. Things look so good that to find a whole skeleton is almost unavoidable but one is liable to be disappointed. But if it were whole, the rest of it!! My!!!

Thursday, August 19, 1909

Went to the brontosaurus today. Mr. Goodrich and Mr. Neal went to town, Mr. Goodrich drove his team. I rode Penny, Frank's horse, and when we went to a spring for a drink, the horse went home. I dug for a while but found it very hard to work as I had tired my muscles yesterday. Found that the bone I struck yesterday is undoubtedly the ischium. Have not found the femur yet.

The construction of a road to the dinosaur does not seem so difficult. That difficultly lessens but that of getting out the dinosaur in good shape increases. It is going to be a tremendous job. But it will be one of the greatest specimens if it is all there. I hope at least to get a hind foot. If the vertebrae are all in position it will be an astounding job. But there is a way. Of all things I must not injure the specimen by carelessness or want of skill.

Friday, August 20, 1909

Mr. Goodrich went to town and home the day before and did not get back so as I had plenty of fixing up to do here. I did not go up to the brontosaurus. Began fixing a kitchen and dining room so we can get away from the flies. Got it covered with a wagon sheet for the roof.

Moved some of the things which were outdoors into it. In the forenoon I stacked grain for Mr. Neal. He has a lot of grain and hay down and it has been so showery lately that he cannot stack his hay. He is anxious to get it done and wanted to help me in return, so we changed work. Got a stack about

2/3 or 3/4 done. In the evening the storm gathered in the east and we had a heavy rain.

Saturday, August 21, 1909

Mr. Neal helped me with the dinosaur. Have a good deal of the pelvis uncovered, am not sure that we have got to the femur yet. It certainly is a little out of place. Aside from that the prospect holds excellent. It may be that the head has dropped down inside the pelvis.

I can shovel and pick for a while but it tires me. We tried to find a spring, prospected for a while, then came home early.

Sunday, August 22, 1909

Today two loads of people came from Vernal to see the dinosaur and there were several loads from other places. For a time the rocks that never had the impress of a woman's foot, and seldom that of a man's, swarmed with people of all ages. Mothers and grandmothers ascended the steep, almost dangerous, slopes with babies and there were men and women well along in years.

I think some were disappointed in not seeing the whole 60 feet of the animal with the bones all uncovered. Mr. Cook, the Vernal Express man, seemed somewhat disappointed I thought. Several took pictures of the animal. Mr. Thorne, the photographer, was there and I guess all his family.

Monday, August 23, 1909

In the forenoon I helped Mr. Neal stack grain and hay and he is going to help me in return.

In the afternoon my man, or boy, Clarence Neilson came with Mr. Goodrich's team. I let the boy stack hay and I came in and made out my financial report for July. My account book had been left in my trunk at Mack so I could not make it out until my trunk came.

Before night there came up a fierce gale, almost a tornado, and it rained and stopped the haying again.

Thursday, August 24, 1909

This morning Clarence Neilson and I took the team and went up to the dinosaur. Did some digging and found that the femur was there but apparently about a foot out of place as it had dropped down about that far from the ilium. I was mighty glad to see that. Found also fragments of the ribs. I think they are the posterior, floating ribs.

I am more and more convinced that we have a prize.

We worked most of the time on the trail and wagon road getting things ready so we can go up with horses and scraper. We got along faster than I expected.

Friday, August 27, 1909

Mr. McHugh began work for me, that is he went up to look over the ground with me and see how the ground can best be worked. He has had much experience in mining, prospecting etc. and I think he is just the man I want. He thinks that dynamite can be used to advantage. Much can be removed with plow and scraper without powder or pick.

We dug forward and found the spines of two presacral vertebra apparently exactly where they should be.

We have begun using the water out of the river; it is roiley yet but we think it is better and softer than the ditch water.

We go to the specimen, or as near as we can, with team and wagon.

Saturday, August 28, 1909 - Hot

Have been at work in camp all day today. Miss Neal sewed the muslin strips together for the side of our dining room and I put them up. Also put up mosquito netting at the end in a temporary manner. I made a table that will seat about 6 and room for dishes etc. Made a wooden bench also that will hold flour.

Made a little cupboard of a box for my letters, writing material and books so things will not be scattered in confusion all over the tent. We have the big tent pitched now. This gives us lots of room now and things are in much better shape. I am in the new dining room now, 9:00 p.m. It is much cooler than the tent. I think when Pearl and Gawin come they can be quite comfortable.

Mention of Mother and me coming brings memories of Father and Mother discussing it years later. She and I had been spending the summer in Missoula, Montana with her mother. The plan was that we would return to Pittsburgh as soon as the summer's fossil collecting expedition was over, which would normally be about the last of September. When Mother heard of the dinosaur discovery and that my father would be late returning she insisted on coming to Utah. Because it was such a wild country, and because of the primitive living conditions and hardships that would be encountered, he tried to discourage her, but it was to no avail.

Sunday, August 29, 1909
A lot of people came from Vernal again. I think six teams in all drove up to see the Dinosaur. Mr. Neal's son and family from Vernal were here, also Mr. Leslie Ashton and wife. I went up with them, explained the thing, answered questions and took pictures. The strong, the lame, the fat and the lean went up. One girl with a wooden leg and a man with only one leg, went up the steep slopes.

Monday, August 30, 1909
Was at work nearly all day fixing up camp. Have dining room enclosed but haven't fixed the doors so we can keep the flies out when the wind blows. They come in swarms.
I hope I can get my crew together soon and get down to business. I am ready to go to work in earnest now.

Tuesday, August 31, 1909

Have not felt so well this afternoon. Slept some of the time. I think if I would properly regulate my diet I would be well.

Went out north of here on horseback to try and find my crowbar. Did not enjoy the ride very much as I had a touch of one of my spells.

Got home about dusk. Felt that I wanted something hot and a cup of tea. Thought oyster soup would be the thing. Had quite a time getting a fire started as the wood was wet and too dark to find some dry wood. Got a little angry but finally got a good fire, put water in one kettle and oyster broth in another. I can't exactly explain it but there is something that makes a fellow happy when he is hungry and getting something to eat in camp.

My boy has not returned with team.

Wednesday, September 1, 1909

It was wet and showery in the forenoon. I was surprised to see three carriages or buggies drive up to the gate. A gentleman came and said that there were two professors in the party and they were anxious to see the Dinosaur. I consented to go up with them. Part of the people were from Vernal and the greater number from Provo. Professor Hinkley was from the Brigham Young College. They nearly, or quite all, seemed to be intelligent people and appreciated the worth of the thing. Professor Hinkley was especially enthusiastic.

Thursday, September 2, 1909

Today we began work on the dinosaur with two men besides myself, Clarence Neilson and James McHugh. Mr. Joseph Hains came after we got up there so we have four men now.

We worked some on the crosscut to see whether neck and head are there. Got into pretty hard rock and our picks wore dull. The ground got dried enough so we could go to work on the trail and did so. We worked at that until noon. Got along quite nicely after lunch.

Friday, September 3, 1909

Today we built a wagon road from the terminus of the road we had used before, up as far as we will be able to go with the wagon. Used picks, shovels, plow and scraper. Got along nicely. We have made all the temporary road except the turn to come back. That is only a few rods. We have a good part of the trail built also.

Some boys were up to see the specimen yesterday and a load of people came up today. The beast has excited a good deal of curiosity.

We could not get Giant Powder at Vernal but can get it at Jensen so about Monday we hope to go at the animal in earnest. I think by the way we built the road we can excavate quite rapidly.

Saturday, September 4, 1909

Mr. McHugh helped Mr. Neal and the rest of us went up and worked on the road. Got it finished as far as we can go with the wagon. ... The trail will need a day or two work yet. We are having trouble getting coal and Giant Powder but we know where we can get them. I hope we will soon be able to spend a full days work on the brontosaurus.

I found it very interesting work making a road though it wasn't easy. There is a fascination in doing things, making the rough places smooth and the crooked places straight.

A load of people, Murrays and Whites were up to view the Dinosaur.

We brought home the loose portions of the tail.

Sunday, September 5, 1909

Was rainy last night and has been cloudy today. The wind blew from the west, then the east, then the west again. So much rain is exceptional for this country this time of year. It is injuring lots of hay and probably grain.

Two or three loads of people came to see the brontosaurus.

My men, Jim McHugh, Joe Ainge, and Clarence Neilson and Mr. Neal were out sitting and lying on the ground a couple of nights ago and talking about fishing and hunting and the wild ducks and geese near here. My desire for a camera for photographing game birds seized me again and I tried to think of some way of getting one but couldn't.

Tuesday, September 7, 1909

Received Giant Powder today, caps and coal for sharpening picks. We have the trail now so we can drive horses up and hope to go to work at the Dinosaur in earnest tomorrow. Hope there will not be many more hindrances. Hope to push the work rapidly.

Wednesday, September 8, 1909

Today we went up the trail with the horses. Worked a good part of the time on the trail getting the steeper slopes reduced but worked quite a little on the Dinosaur. Some of it is extremely frail. It will require great care and skill to get out right. Used some plaster today to protect frailest portions.

It is doubtful, according to Stewart, whether I can get money to keep at the thing until late, though I think I can. I have to do the cooking but Mr. Ainge is good to help me. I enjoy getting supper. It is the pleasant part of the day. The weather bids to be pleasant now. Flies and mosquitoes have bothered much.

Thursday, September 9, 1909

Today we first tried Giant Powder and it certainly does the work and loosens up a lot of ground. I did not think it would work so well in that stuff. The plowing and scraping too are moving along splendidly. I am well pleased with the progress of the work.

Sunday, September 12, 1909

Mrs. Neal said when she went to Vernal that she would get Pearl if she came to Vernal but I thought Pearl might not get there Saturday evening so I decided I would drive to Vernal. I met Mrs. Neal on the road and she said Pearl did not get there yet.

I went to Vernal, put the horses in the livery stable, went to the hotel and got the hat I left there. Found after a while that there was a telegram. It announced that Pearl would start from Mack Sunday. I got her on the telephone and certainly was glad to hear her voice. I arranged to meet her at Alhandra (a ferry across the river) on the Green River. You bet I was anxious for the next day to come.

Monday, September 13, 1909

Got my business done in Vernal and got started for Alhandra a little between 12 and 1 o'clock. Drove down to Green River with high hopes. Got there about 3 p.m. Had to wait a long time, at least it seemed so to me. Finally I saw the stage come down the hill. When they came back from the station to the boat I was glad enough to see Pearl wave her hand. A man took Gawin into the ferry boat. Soon we were on our way toward camp happy as two lovers are when they get together after a long separation. We got to the Neal Ranch just about dark.

The stage coach and freight wagons coming from Dragon to Vernal traveled a route that is not used today. It was almost a direct line from Dragon to Vernal. The Green River was crossed by ferry boat at a point known as Alhandra, about eight or ten miles below Jensen. To get us Father first went to Vernal and then to Alhandra. After getting us from the ferry he went up the west side of the Green River to the Neal Ranch instead of going back to Vernal which was several miles out of the way. This saved considerable time and enabled us to arrive in camp that night instead of sometime the next day if he had awaited our arrival in Vernal.

Sunday, September 19, 1909

Have not been feeling well for several days on account of a bad cold. We got up a little later than usual. I shaved and got ready to go up to the brontosaurus, as I fully expected to see a good many people from Vernal. They came but, as usual, not the ones I expected. Five or six teams went up. Some were from Vernal, some from Salt Lake City. I knew only one or two. Pearl and I rode up with Mr. and Mrs. Murray. The people were mostly young. They were much interested and did not seem disappointed. They started on the wrong trail and had a hard time to get back.

Mr. Murray and family took lunch with us in our little dining tent.

Monday, September 20, 1909

Hot, as usual, from 10 or 11 until 4 or 5 o'clock. Pearl is cooking for us now. When she dressed Gawin this morning she put a sweater and overalls on him. He looked like quite a boy.

While going to work or just before I started, Mr. Neal handed me a copy of The Editor, the first one I have received since leaving Pittsburgh. I thoroughly enjoyed reading it and as usual it gave me a new inspiration and a new hope. The most vitally interesting thing is that the Bohemian Magazine is to get out of the old level and give us something new, original, unconventional. I intend to subscribe. I long for something new – new lines of thought. I am sick of the "good stuff" the "good for nothing stuff".

Was very busy today on the hill; worked at two or three places at about the same level as the brontosaurus skeleton near our trail. One is apparently the toe bones of a very large dinosaur and the other a large part of a skeleton of a small one. I think both are well worth saving.

Tuesday, September 21, 1909

Windy last night and today. Wind from N. W. today and cold. Pearl and Gawin went up with us today. It was cold but they enjoyed it. Mr. McHugh drilled and made two shots moving a lot of hard dirt as usual. Have found some more bones, three look like ribs.

Wednesday, September 22, 1909
Cold last night. Wind from N. W. nearly all night. Mr. Neal said there was frost. Made some notes this morning going along the road to the dinosaur.

Small golden-rod like flower in bloom & fading, greasewood faded, cottonwood trees dark green turning yellow, "Lone Tree" mostly dark green with faintest tint of yellow. Streaks of light toward sun across river. Light sparkling like diamonds on water this side. Beautiful trees in light and shade on sandy shore on other side of river. Fleecy clouds above Blue Mountain, in clear cold light blue sky. Contrast between this and dark, hazy, long, flat-topped ridge. Joe singing. Jim started humming or singing - spitting, whistles. Clarence, elbow on leg, whip in hand which is against face, humming and driving. Brown mare going along about business. Gray trying to keep up. Little, stunted sunflowers in bloom a few inches high.

Thursday, September 23, 1909
A fine clear day. Cool last night and this morning. I got up started fires in the tent and under the grate and got most of the breakfast. We have stove in dining tent now and mostly enclosed. Have sand on the floor so it is quite cozy and comfortable. Pearl has our sleeping tent quite comfortable too but the little oil stove is hardly enough to warm it when it is cold.

This morning the men had drills and picks to sharpen and I worked around camp and began packing fossils. We haven't one box packed yet.

The brontosaurus is more disturbed than I thought it was at first, but how much I cannot tell. It seems going more nearly straight down now. Received a letter from Stewart yesterday. He says to go ahead with it.

Friday, September 24, 1909

Morning wagon notes. Sky-blue, cloudless. "The sky is blue" is an old, old saying probably a hundred times older than history but to me it grows to mean more and more and all the colors of the spectrum are in it. Blue Mountain is dark with bluish haze, cliffs, gorges are not well-defined. On the island in Green River light is shining through tall slender trees and underbrush.

Still want to read something I cannot get. Not man's outer life but inner and his connection with nature. Some unusual story. Am going to try Bohemian Magazine. Men long for something dramatic, some crisis always arouses interest. I often thought I would write my views of the dramatic in religion. I suppose it is a reaction against the humdrum way of life and it may be an incentive to higher things. People tire of common life.

Saturday, September 25, 1909

Worked on the brontosaurus until about noon. Ate lunch and went to the hills further north to get some wood. We pulled off the cedars with a horse and snaked it down a hill, partly on a ridge.

Found a peculiar nest, at least it looks like a nest, but I don't see how any bird could lay such big sticks in such a position.

We quit before four o'clock as I had to go to Jensen and get supplies. Pearl went with me. I had never been in Jensen before. It is quite pleasantly located on the bank of the Green River by the ferry and there is a fine view of Split Mountain, Green River Gorge.

The dry cedar wood from the hills to the north of the discovery was the principal fuel used for starting fires and cooking during all the early years of excavation at Dinosaur. The nest mentioned was later found to be an Indian burial. The Indians in this part of the country quite often buried their dead in cedar trees by using sticks woven together with strips of cedar bark. The sticks apparently remained intact for years after the bones had disappeared.

Sunday, September 26, 1909
We spent the day as we have often wished to spend Sunday.
Yesterday I got the books which I had ordered from Sears Roebuck. They are Ayesha, The Blazed Trail, Brewster's Millions, The Young Explorers, The River Boys on the River, Frank the Young Naturalist and Robinson Caruso.
We went down to the cottonwood grove S.E. of here, by the river, in the forenoon and read Ayesha. It seemed about the thing I had been wanting. I always enjoy Haggard. After dinner read The River Boys on the River, thought it would be an appropriate place but didn't read much about a river. It was mostly, hazing, scoffing, stealing and playing tricks. I don't enjoy such stuff.

Monday, September 27, 1909
The boys have fixed a drag with a box and today Joe used that to haul rock on. He likes it much better than the scraper. We got along nicely today.
I don't know how much longer we will be permitted to stay. Expenses are high now and I presume we may have to quit about the middle of October especially if we cannot get our tickets extended.

Friday, October 1, 1909
At work as usual. Did some more work on the trail. The days are much alike now. The rides to the work are usually enjoyable.
We usually get up at near six o'clock. I go and quickly start the fires. Pearl gets up and usually Gawin wakes up so she has to dress him. I help get breakfast which consists usually of three or four of the following: Pork, ham, bacon, eggs,

hot biscuits, Johnny Cake Corn Fritters, Cream of Wheat, sometimes pancakes, coffee, bread and butter.

After breakfast the men go out to the forge and sharpen the picks and drills. I do some work about camp, packing specimens etc. Pearl puts up lunch. Clarence drives up and we put lunch and water bag in. At river we stop to water horses and get water. When we get to fossil grounds Jim usually goes to drilling and the others to hauling, shoveling and picking. I tend to the work near the skeleton.

Monday, October 4, 1909

Took supper at Mr. Neal's. When we went to go to bed in the tent there came up one of the most terrible flashes of lightening and the worst compound thunder shocks I ever remember hearing. The water went in a kind of a mist through the tent and made it a little damp inside so Pearl took Gawin and went in the house to sleep.

Wednesday, October 6, 1909

Have been at work getting ready to take things up. Want to take up the foot - will take up all but the large skeleton and as much of that as I can. Don't hear from Stewart yet and haven't money to settle my bills. Am hoping to hear from him every day.

Am sad this evening. Have a letter from Ida. It looks as if the tumor is growing again. The doctor thinks probably it is. Alas for us poor mortals. Though I can enjoy as deeply as anyone and there is much that is sweet and beautiful yet I fear I am getting to be a terrible pessimist. It is hard to establish a philosophy that is optimistic and yet will fit the terrors of this world.

Thursday, October 7, 1909

Rejoiced to get a letter from Douglas Stewart with certificate of deposit of $350 for field expenses. Have now received $1000 in all. I estimated that it would take $2000 to $5000 to get the brontosaurs out. It has cost now between $500 & $600 and we have a lot of work done. Trail and road built and dirt removed from a good part of one side. Also lots of camp supplies, tent, etc. It will cost in the neighborhood of $350 per month.

Friday, October 8, 1909
   Got ready to go to Vernal as we needed many things and I wanted to put money in the Vernal Bank. It is worse than digging, a good deal, to ride to Vernal in a lumber wagon. The roads up Brush Creek are very rough. Got into Vernal between 1 & 2 p.m. Put horses in stable and "flew around" trying to get my business done up.

Saturday, October 9, 1909
   Still in Vernal trying to get my business done up. Tried my best to get lumber but there was none to be had. Could not get plaster either but found a man who is going to try to make some. Saw Mr. Johnson and he said I could get lumber at the Griffin Mill. Got started about two and got home about half past six. I had most of the male clerks at the co-op waiting on me at one time or another and they had to go nearly all over Vernal to all the stores to get what I wanted.
   Mr. T. M. Young asked me if I knew where they could get a principal for the Jensen School (He found my wife was a teacher). I said she might if it were not for the baby. I didn't think she could manage that.

Sunday, October 10, 1909
   In camp all day. In the forenoon I was fixing up and we and Mrs. Neal were trying to decide the school proposition.
   I got an old box or crate that I got in Vernal, put some shelves in it and put it at the end of the table in the dining room for a book case, and fixed up other things. We have a pretty neat camp.

> Took a quilt and went down in the corner of the orchard by the plum grove. Took Ayesha. Went to sleep, all three of us, and woke up cold.

Mention of the orchard brings back memories of the Neal Ranch, one of the nicest ranches in the valley in those days. It was the last ranch up the Green River from Jensen on the west side. A ditch from Brush Creek furnished water for irrigation. Excellent hay and grain, which were raised in abundance. A large log house stood on the lower side of the ditch. Back of the house was one of the finest orchards in the country with several varieties of apple, peach, pear and apricot trees, a plum grove, and a wonderful raspberry patch. Adjacent to this was a large garden plot with corn, potatoes, melons, squash and practically everything that grows in a garden. Above the ditch and next to the road were the corrals, stable and pens for the livestock and poultry. My father's camp was near the orchard and not far from the house (I'm not sure whether it was above or below the ditch). If it had not been for this ranch and the Mr. and Mrs. William Neal family, the development of the Dinosaur Quarry would have been a great deal more difficult. Not only during the early work at the quarry but through later years, the Neal Ranch was a place to get help in an emergency and badly needed supplies, food, vegetables, meat, hay and grain. Anything Mr. or Mrs. Neal had on the ranch was ours to use without restriction. My father and mother were indeed fortunate that such a ranch and such people were as close to the dinosaur deposits as they were. It was slightly over two and one half miles from the ranch to the quarry.

> Monday, October 11, 1909
> Went up to the dinosaur again this morning. The boys had been working around the three specimens about as much as they dared to. Took my gun along. Got some shot gun shells yesterday. Killed three of the chipmunks that have been pestering us so by eating paste off from the specimens.
> Had some excitement again today. I had given up finding the skull anywhere near in place as the cervicals suddenly

come to an end. In digging around and above, found bones convex and flattened where I did not expect to. I did not know but it might be a skull. The boys were interested apparently as much as I. It was an exciting time. I finally concluded that it was not a skull but it looked like sides of centra of vertebrae; dug further. It was and looked as if wrong side up. Dug around the edge and the end was very concave. The concavity, I thought was behind in these dinosaurs that would make the vertebrae turned the other way and the head ought to be toward the body, neck doubled back on itself. Was the whole neck wrong side up on the part turned back? Skull looks certain again.

To one unfamiliar with dinosaur bones a little explanation concerning this first skeleton should be interesting. In taking a quick glance at a photograph of a portion of the charted bones it would be possible for one to assume that the spinal column of the upper skeleton (the one first discovered) was pretty much in natural position with the head and a portion of the tail missing. This was, of course, what my father was assuming. One must remember that everything below the horizontal line drawn on the picture was either below ground or in solid rock at the time of the discovery. Also in most fossil collecting only a small portion of the bone is uncovered in the field; only enough to determine the position of the bone so that it can be blocked out in the rock. The rest of the rock is removed in the laboratory when it is prepared for mounting.

On closer examination of the chart one will notice that the dorsal, or back bone section, with the high dorsal spines which should be immediately ahead of the pelvis, is not there but is lower down. In its place, butted right into the pelvis and in almost perfect line with the sacral and tail bones, is the neck turned upside down and end for end. What was thought to be a ribcage, which were found soon after the discovery in the position in which they were expected to be, were the ends of the spinal processes on the huge neck bone of the brontosaurus. Most of the leg and foot bones, both front and hind although scattered, were found, as was the case with the ribs.

The tail, part of which was at first believed to be missing, although somewhat disturbed, was complete. Except for the skull the skeleton was practically complete and now stands mounted in the Carnegie Museum (There have been changes to the display at Carnegie Museum since Gawin visited and I'm not sure which dinosaurs are exhibited in the big hall at this time, DDI). The ancient river that is thought to have deposited this skeleton was quite likely the one responsible for the strange manipulation of these bones. This all caused a great deal of anxiety, and probably some fretful nights sleep for my father, as is indicated in some of the diary entries. Such are some of the problems of the fossil collector.

>Tuesday, October 12, 1909
>
>We got up a little earlier than usual; saw a sheep wagon coming and wondered where it was going. Soon they said it was the school wagon coming for Pearl. She did not expect it so soon but they waited for her. It seemed queer that she should come out here to be with me and take care of camp and should take the principalship of a school. It was quite an eventful day.

>Thursday, October 14, 1909
>
>Working to get in front of brontosaurus. Everywhere we tried today we would come up against bones and I would have to treat and dig around them and that was about the only place we could work.
>
>I was digging around the pelvis of the small dinosaur above the large one's neck where there is an ilium, ischium, pubis, femur and tibia, I saw some bone to the left and near the ilium. In digging below it I broke off a slab and uttered an exclamation of surprise and delight. It uncovered so I could see parts of five of the vertebrae in succession and beautifully articulated in position. It is almost as interesting to me as the big one.

Saturday, October 16, 1909

Still trying to get into the ledge in front of the dinosaur but bones in the way. They were evidently not from the skeleton.

I have longed for weeks to go to what I have called Sandy Island, a little way above the Indian Ford. Joe went up to Johnnie Miller's camp and got one of Thorne's boats and brought it down to the bank opposite the island so I could visit the island Sunday.

The Indian Ford was a well known river crossing at the upper end of the Neal ranch. It was used by the Indians and I can remember it as being one of the few places between Jensen and Spilt Mountain Canyon that the river could be crossed in a wagon at low water. It was the main ford for crossing cattle or crossing by horseback. Father Escalante was supposed to have crossed the river here in 1776 and this is where the Escalante Monument now stands near the road to Dinosaur National Monument. A few years ago the ford, or riffle, disappeared and it now seems to be one of the deep places in the river.

Monday, October 18, 1909

Would be very happy now were it not for Ida and Nettie. The suffering of humanity gives a dark background to everything and when I am successful and the prospect looks bright their sad lot comes up and I feel I have no right to be happy.

As far as success in my work is concerned, I could not ask for more. For years I have wanted to get nearly complete skeletons. I had found much that was scientifically interesting but felt I had not found such complete things as I wanted to and others were ahead in this respect & there is no skeleton in the museum set up that I have collected. Now I have suddenly most of the skeleton of a large dinosaur, apparently the most complete that has been found, and now there is a good prospect of a little fellow which is probably new.

Then there is a good part of another skeleton not so nicely preserved and a fine foot.

Friday, November 5, 1909
Work getting especially interesting. Some of the bones puzzling, especially neck. Working around tail and pelvis of small dinosaur to get at part of neck of larger one, or ones.

Saturday, November 6, 1909
Pearl and Gawin went with us up to the Dinosaur. It is getting mighty interesting now. I have changed my mind again about the neck and think it is the anterior portion that is under the back of the other. I think that there are two large dinosaurs there instead of one but am not sure. If not there is surely nearly a whole skeleton of one, at least the back bone.
We got down the tail, pelvis and femur of the little dinosaur in a block. Joe made a box and we put it in. This is the first large specimen to come down. It makes us realize that it is going to take a lot of lumber. Some boxes will probably take 150 to 200 ft. of lumber.

Monday, November 8, 1909
Yesterday was a red-letter day for us. I received papers on dinosaurs that I have long been wishing for. Got a letter from Raymond who said I had been appointed lecturer in vertebrate paleontology in the University of Pittsburgh and that he had read my article in The Columbian. I suppose the latter must be "The Devil's Play Ground", which I sent to the Burnell Syndicate last spring. Probably they have written me and the letter was miscarried. In the university there is no teaching in our lines as yet. Raymond is lecturer in invertebrate paleontology but it will be an honor to us and we will get in on the ground floor of a growing university. I had longed for some such thing but the realization seemed doubtful and remote.

Brought down the toe with the part of the small dinosaur, the first large box we have brought down. The drag and snake road worked all right.

Tuesday, November 9, 1909

Charley Neal came out from Vernal with his mother. His father and I got things ready and we started for Split Mountain Canyon. Got started between 8 and 9 o'clock. Put team in cave which made a nice stable. Pitched the tent in a little enclosure built by prospectors. Got dinner and started up the canyon. We had a boat along and went across the river; found we could not row up. Went across a rocky and sandy river flat then up over a rough, steep slope. Found shells of brachiopods and bryozoans and one trilobite. Stopped and made a collection. We then went on to where Charley had found some fossils. After a while came down to a river flat again. I left the specimens near a bend in the river; found little pools with beautiful reflections. Quite a distance further up found another horizon of fossils, Devonian probably. (editor's note: Almost certainly Mississippian or Pennsylvanian as there is no Devonian rock known from this part of the Uinta Mountains, SAB)

I saw it was getting dark. It was cloudy and looked at my watch. It was nearly 5 o'clock and a rough trail behind. Charley Neal was way ahead, out of sight. Mr. Neal was still going. I started for camp building fires along the trail; managed to get back to the river. Thought once I might not be able to make it. Lighted an old cottonwood tree once. After some timid hesitation I crossed the river after hoping they would come. Was with team and felt safe only thought I would have to go back after them. Built a fire in cave where the horses were then went to tent and built another. It was dark as it could be but could see white tent. After I had about given up I heard loud shouts across the river and knew they were safe. Went over and got them. In coming across rapids my hands were on the edge of the boat on both sides and fingers in the water. It was not a very pleasant ride.

The next day it rained pretty hard in the morning. We debated what to do. I rather wanted to go up the canyon again but Charley was anxious to get home. Finally decided to go down river with boat. I found it a larger run than I expected; shot a good many rapids. A new experience to me and I learned some geology along the route. Did not get home until after 5 o'clock. Stopped at dredge for lunch.

⌘ ⌘ ⌘

29

DIGGING OUT DINOSAURS:
A Cold Winter

Up until this time the work on the newly-discovered dinosaurs had been pursued on a day-to-day basis. The objective was to obtain as much information as possible as to the extent and importance of the discovery with the idea that their endeavors would soon be terminated because of the lack of finances. There was, perhaps, a faint hope or dream that by some miracle this might not be the case and the work might be continued. The diary does not make clear just what happened but sometime between the 9th and 13th of November word was received to continue the work through the ensuing winter. Andrew Carnegie had become extremely interested in the find and personally added $5,000 to the $10,000 paleontological fund of the museum. As the diary continues, after a brief lull, all were elated over the news. Hasty preparations were made to establish a new camp near the site of the excavation before the approaching winter, that might set in at any time. The new camp site, which would eliminate the long daily drives through the cold, was considered an absolute necessity.

> Saturday, November 13 1909 - Written the 14th
> Went up to the gulch this side of Dinosaur to see about building a road up and making camp there. Found everything favorable. Got up to where we wish to camp with wagon, and cleared ground for tents, wood pile, etc.

> Written the 22nd - Have been so busy that I did not have time to write the last week so will have to write from memory.

We found it was not so difficult as I thought to build a road; only two or three places that needed much work. One where we enter the ravine through a ledge of sandstone and cross the little stream that flows through there now. Another was across a little side ravine where quite a little cutting and filling were needed and the last one was across a little side gulch and by a ledge of rocks. We found that the rocks pried off easily to fill and make the road wider.

All were jubilant over making camp for ourselves and by ourselves. We planned where we would put the dining tent, sleeping tent, wood pile and stables. All was cozy and nice. Then, as everyday since we have been at the ravine, I have wished for a camera - a Graflex.

I noticed sage brush, buffalo berry, squaw bush, willow, alder or birch, clematis, plume grass, box elder, cottonwood, cactus. Wanted to study the plants and animals of the ravine.

Sunday, I think it was, I determined that I would go to Vernal and have things straightened up and get ready to work without hindrance before I came back. I had expected money every day for two weeks or more. I had got where I could hardly work and bills were not settled.

Before transcribing my father's diaries and notes I had always assumed he approached the dinosaur discovery from the gulch to the west of the visitor center of Dinosaur National Monument, as the present road does. This was not the case, however. He approached it from the gulch to the east and soon after the discovery a road was built up this gulch, as has already been mentioned. They continued to use this road until it was decided to make a new camp and work through the winter.

The gulch to the east was a dry gulch and had little desirable to offer for a camp while the gulch to the west, although not as accessible, contained an almost ideal spot for a camp, and was later named Camp Gulch. There was a nice spring with trees adjacent to a small flat surrounded by hills that made it about as protected from the elements as any place in the area could be. All this, together with an abundance of dry cedar wood in the sand hills to the north and its proximity to the quarry site, made it as close to ideal as possible in such a wild and barren country.

Monday, November 15, 1909
Got Mr. Neal's bay mare, Johnson's old gray horse and Neal's light buggy and started for Vernal. When I got there one of the first things I did was telegraph for $500 and then I tried to set other things in motion. Tried to hear from (locate) my barrel of alcohol which I had telephoned to Grand Junction for October 22. It was found in Dragon. Mrs. Douglass went with me to get her teeth fixed and do some winter shopping. Mr. Neal went to Vernal the same day with a load of wheat.

Tuesday, November 16, 1909
Mrs. Logan did not get the telegram off until today. I did not know when I would return and Pearl was anxious about the baby so she thought she would go home with Mr. Neal.

We wanted to get a forge and stove. I had fears about the former but not about the latter. Went to the store on the corner where they had them when I was in town before but they had no camp (sheepherder) stoves left. Ashton's had not one left. Went to Johnson's and they had one, not just the kind I intended to get but was glad to get it.

Leslie Ashton had a portable forge that he had ordered for another man. He tried several times to get him on the telephone but could not so said I could have it and he would get the other man another one. Then I had to have anvil, tongs and hammer. I got never-slip horseshoes and shoeing outfit and a hundred and one other things that we needed.

In the evening I took dinner with Mr. F. M. Young, Superintendent of Schools. He appears to be a warm friend and is always enthusiastic over our work.

Wednesday I was busy most of the day getting supplies. Was hoping every hour I would get a telegram telling how much was deposited. I was bargaining for about $150 worth of goods and little or no money in the bank account.

Thursday, November 18, 1909

Was further delayed on account of not getting the telegram. Got things as nearly ready as I could and finally just before dinner it came. I hastened to settle bills and get things loaded. It was 2 o'clock, however, before I got things loaded. Then in Ashley Creek bottom I found I had lost some stove lids and went back and hunted (got one found two in wagon). Then a man stuck on a hill detained me. I think I got home about 7 p.m. I had a pretty good load and felt my trip and stay had accomplished something and that our camp was pretty well started. I found a man who was selling beef by the quarter so I got a hind quarter.

Pearl was tickled to have me home and was jubilant with the prospect of moving camp and having a stove. I think all were pleased except Mr. and Mrs. Neal. All their children are away and they will feel lonesome.

Friday, November 19, 1909

The boys had been busy at camp during my absence. This morning we went to the camp. They had the spring dug out, cellar dug, covered, and vegetables in it, and a good part of the cabin built. I took hold and helped and we got along very nicely. Joe went home, got nails and to Jim's camp to get building paper.

Saturday, November 20, 1909

Moved our sleeping tent in the morning; got it well toward done and I returned to tear down the dining tent and get Pearl. When we returned the sleeping tent was up, floor laid and the stove in the dining tent. Pearl was much pleased with camp.

Clarence and I went back to get another load and it was getting dark when we returned. It was unusually warm but this was the only indication of a storm I could perceive. Soon after returning it clouded up and began to sprinkle. We just got things well enclosed, tents fixed so they wouldn't leak. It rained some the greater part of the night and part of the next day. It doesn't get muddy here; it is all sand. We have nice clear spring water a few steps from the door and there is lots of dry cedar on the hills a little above camp.

Sunday, November 21, 1909
We worked all day to get things straightened up in camp, all of us. Joe went home and brought up the top of a sheep wagon for sleeping room. Succeeded in getting better prepared for the next night. All were apparently glad to get there and we talk a good deal about how nice the camp is and how much better than being down there. The whole thing seems like a romance.

Monday, November 22, 1909
It is nearly eleven o'clock and I am sleepy. Came home tired tonight.

Joe and Clarence took the forge and grindstone and other things around and up to the diggings. I had to come down in the forenoon and it took me three minutes. Of course it will take longer to go up.

Sent $40 to pay on a Graflex camera, part of it to be paid in December and the balance Jan. 10th if they send it, but it is very doubtful indeed.

Wednesday, November 24, 1909
Got up at 5 o'clock, that is I got up and built a fire in the tent. Went back to bed until it got warm then came into the cabin. Started a fire and mixed pancakes. Pearl soon came in and helped.

Joe went to his home to get timbers for tripod. He got posts for a clothesline also and part of the lumber that we left at Mr. Neal's.

Jim made a fire in the forge and sharpened a lot of tools. Mostly all were dull. It was fine to have a forge and grindstone. Jim then went to work at the tool room in the cliff. Clarence and I worked at the Dinosaurs.

Friday, November 26, 1909

At work at the Dinosaur. It was not a very pleasant day.

Soon after we came down to lunch Mr. Neal, Mr. and Mrs. Morrison, Frank and Ollie came. We had to eat lunch, or I would, as the people and school children were coming up from Jensen and I thought the teachers would be there. Professor Buckley had a break-down and did not get there for some time. Probably there were 50 or more people there today.

In the afternoon the wind blew a gale.

Monday, November 29, 1909

Cloudy the greater part of the day. Jim and Joe went up to the diggings while Clarence and I went to Neals' and Murrays'. I wanted to see about some pigs and a cow. We have lots to feed a pig and it will cost practically nothing.

We got some hay. Mr. Neal asks $6.00 per ton that I think is quite reasonable. We will need considerable hay if we get a cow especially.

In the afternoon I went up to the Dinosaur. It was unpleasant but I managed to do some work. I must stay with the bones now every day I can. Since we began moving the work has gone slowly there.

Wednesday, December 1, 1909

The boys, Joe and Clarence, worked at the Dinosaur in the afternoon; hauled hay in the forenoon, one load. Jim working at the tool room in the rocks. I did not know it was so hard a job or I might not have undertaken it. It will pay for itself though I think. We can keep water without freezing, keep powder so it will not be so hard to thaw out. Can keep tools away from the rain and snow, have a place to go in when it is too cold, to warm up.

The tool room was a short tunnel blasted into the cliff at about ground level at the time it was constructed. It was boarded up in front and had a small wood stove in it. Later, when the main cut was run through the hill to expose the bone layer, it was left some twenty feet above the bottom of the cut and was reached by a ladder. It was used all through the early years as a powder room and a place to store plaster. These supplies had to be transported across the cut on a cable and trolley. Later another similar room was excavated in the east end of the quarry for a tool room.

> Thursday, December 2, 1909
> Snowed the night before and a good part of the day.
> The boys got hay in the forenoon and went to Murrays' and got a cow and pig in the afternoon. Paid, or am to pay, $5 for the pig and get the cow for keeping it.
> Some time ago I sent a check for $40 to pay on a 3A Graflex camera, the rest to be paid in December and January if they wanted to do that. They have consented and the camera should be on its way.

> Saturday, December 4, 1909
> Cold but bright and pleasant at camp and diggings. All of us worked at the Dinosaur most of the day. The first time all have worked for some time. Joe and I quit early and came down to make a door to the sleeping tent line on the south side of the cabin etc. Finished the pig pen and put the pig in it. Dug it in the side of the bank and made the pen of birch, just above here on the gulch. The cow and horses have no stable yet. Started yesterday to dig the cow stall in the bank.

We now have nearly all the comforts of home. The cow will give plenty of milk for our use and the pig eats the slop we would have to throw away and which would contaminate and disfigure our home.

It was cold the night before, we kept a fire all night. I got up two or three times in the night to put in wood and coal.

Thursday, December 9, 1909

Stormy - Can't wake up at the right time lately; am either too early or too late.

Clarence went to Murrays' this forenoon. He got six ducks and we killed and dressed them this evening. Got some nice feathers as well as meat. The boys came in this evening and we picked the ducks. Afterward we had coffee, cookies and raspberries and chocolates. All were fine and we enjoyed them.

Friday, December 10, 1909

Cold. Quit storming. A nice bright day. All working at Dinosaur. Finished the cave and moved into it. It is none too large but we have a stove, forge, grindstone and water barrel in it.

Joe went to Jensen in the afternoon; took letters and brought back mail, a good deal of it. One letter was from Dr. Holland. As requested, he deposited $500 in December so they deposited $1,000 within three weeks or so. That makes $3,000 in all I have received. I knew it would cost a lot of money but I am not disappointed.

Was writing checks and letters and straightening up business until late.

Saturday, December 11, 1909

Jim and I worked at the Dinosaur. In the forenoon Joe and Clarence snaked down wood. In the afternoon they shoed horses.

We are striking new bone all the time.

Received a letter and check from the Burnell Syndicate. The check was for "The Devil's Play Ground." I did not expect to get less than $40 or $50 . It sold for $20, I suppose, for I got $18. Well that is better than nothing and I guess it paid for the pictures.

Tuesday, December 14, 1909
Still very cold, especially nights. It is not uncomfortable, however, as there is little wind. Was very frosty this morning and vapor arose from the open places in the river.

We had breakfast at about 7 o'clock. Got up to the diggings earlier and accomplished more.

This is a day to be remembered as we took down the first really big specimen - part of the neck. I was somewhat surprised as it seems that we have only one vertebrae of the neck instead of two as I supposed. With our pulleys anchored in the cliff above three of us eased the thing down while Jim loosened it and eased it down with a crow bar. It came where I wanted it and right side up, some of the bones were broken where the block parted but that had to be.

Wednesday, December 15, 1909
Seems to get colder and colder every night. It was snapping cold last night but we do not suffer much with the cold because it is so still.

We quit about a half hour before five as it got so cold after the sun went down.

Thursday, December 16, 1909
It was cold and stormy this morning. Began to snow or frost and turned into snow. It would have been difficult for all of us to work at the quarry so I thought it would be a good time to try to go to the east end of Sandy Island and get some timber for stable and anvil block. We went down and tried the ice and I saw it was not safe and was going to give it up as I did not want to take a bit of a risk. We concluded to try

it a little higher up and got across in safety. Found just what we needed for an anvil block and cut a lot of timber for the stable.

I enjoyed getting ahold of an axe again and felling trees. It seemed a little like old times. There is probably good dry wood enough on the island to last all winter.

Saturday, December 18, 1909
Last night was the coldest we have had but I think tonight will be about as cold.

Sent Joe Ainge with the team to Jensen to get mail express, and provisions. It was a cold day and part of the time none too comfortable. It seemed I could not do much as everywhere bones needed a shellacking, plastering and pasting and it is too cold to use paste and plaster. So the weather hindered us some but I think it will moderate soon.

Got my Graflex camera, a thing I have wanted for so long. For a year or more I have been "crazy" to take pictures of wild animals, especially game birds.

Monday, December 20, 1909
Snowed the latter part of the night and this morning. Did not go up to the Dinosaur in the a.m. I think we better get a shelter for the bones if we are going to have a winter like this. It has been down to –32 or –34... –30 to –38 at Ouray they say.

Friday, December 24, 1909
Last night undoubtedly was the coldest night we have had. I do not think it was much less than –40. Tonight will, I think, be colder. There is not much breeze, however, and it is not hard to stand, though most of the time there is frost in the atmosphere.

Joe and I worked around here. It seems that now we have things fixed up so we can spend our time at the Dinosaurs. We have a good stable for the cow, a better shed for the horses than most have in this country, and a comfortable little pen for the pig. The pig is growing like a weed, the cow is doing well and the horses are looking well.

Wednesday, December 29 1909
Received a copy of the Columbian Magazine which is the first issue. It is the first number of a new magazine. It is a little strange. I had an article in the first number of Guide to Nature magazine. Received one dozen rolls of films from Jute's also a new 1910 diary.

Friday, December 31, 1909
Weather moderated. Snowed the greater part of the day.
Well this is the last day of 1909. To me it has been probably as prosperous a year as any I have spent and the latter part of it, especially, has been, or would have been, about the most satisfactory and happy that I have known were it not that sorrows and pains of others cast a shadow over me. Nettie has ever been a source of solicitude and sadness to me, and I have so much wished that she might be well and happy. Now Ida is so much worse and in so dangerous a condition that I almost forget Nettie's trouble and before I worried very little about Ida. Ida begins to feel she will not be with us long.
I do not know whether I have advanced in the highest sense. I wish my mind could be settled. The question of life has become to me a great philosophical problem. We cannot judge whether it is good or bad unless we know about the future. I am pretty pessimistic at times of late and do not see how anyone can help being who looks things squarely in the face, unless they believe in a hereafter.
I have accomplished some things but there is ten times as much that I wish to accomplish. Sometimes I am full of ambition (crazy to accomplish certain things). I especially wish in a side innovation to write books for boys. I am getting, in some ways, back to boyhood.

My ambitions have been satisfied in many ways this year. I discovered a new locality for fossil mammals in the Wasatch and I have finally found the best brontosaurus and a small dinosaur. I did not care to do quarry work but the magnitude and glory of the work has spurred my ambitions and it is full of interest and pleasure.

The spot where we live is now almost ideal; it seems my natural habitat. I have been getting more and more interested in photography and have longed for a Graflex camera. Now I have one and have it paid for.

We have been getting better and better fixed financially. We have been wading in debt since we were married. It has been a struggle working for Carnegie and just getting through by a hair. Now it is easier and we are living a simple life and we both earn something and have our board paid.

Wednesday, January, 5, 1910
We all went up to the diggings in the forenoon. Moved a good deal of dirt. We are beginning down low and preparing to sink down 15 to 20 feet. Now is a good time to do this work, during colder weather when we cannot work at bones.

This evening Pearl took Gawin out doors and soon came in saying an animal was out there and it made a thudding noise. Joe grabbed the gun and went out. Shot at it, only a few feet way but did not hit it. Chased it to the ravine and shot at it again.

Thursday, January 6, 1910
Very cold. We think it is about down to -40 Clear. I did not suffer much working. A fellow has to look out for his ears, nose, and toes. I have a pair of wool-lined shoes and my feet seldom get cold.

We do not very often see anyone but our own party but a young fellow, VanGundy, was up to the Dinosaur yesterday and today.

Saturday, January 8, 1910

Still cold as ever. We are surely getting a winter in earnest.

This morning Clarence and I went to Neals' and got a load of straw, also 3 hens and some buttermilk. We had to watch our extremities to keep them from freezing.

Jim and Joe went up to sharpen some steel, drills, etc. and then came down and spent the rest of the day making an excavation for a place to keep ice.

Tuesday, January 11, 1910

Not quite so cold as it has been but still cold. Joe went this morning to get a load of coal at the coal bank on Brush Creek. Has not returned yet.

Yesterday Joe went to Jensen and got our mail. Raymond sent a clipping from the Gazette Times in which it was stated that Carnegie has provided means for the getting out and setting up of the Dinosaurs. I hope it is so. It would be too hard on the museum and I fear the work would be delayed.

Wednesday, January 12, 1910

Weather about the same as the last few days. Am on the sick list; have a beastly cold. Haven't done much work.

Joe got back about 11 a.m. There were 12 teams ahead of him waiting for coal and he didn't get loaded until after dark. He got 100 lbs. of dynamite. Was mighty glad of that. Got only 1000 lbs. of coal.

Wednesday, January 19, 1910

Cold last night. Judging by my feelings it was about the coldest night of the winter but I can't tell by my feelings.

We all went up to the Dinosaurs today. Jim is tearing thing up but it is not so nice getting away dirt as it has been. We are getting along pretty well though.

Saturday, January 22, 1910

Perhaps not quite so cold. Got along nicely with the work today. Have four nice new picks and they work fine.

Joe stayed at the dance in Jensen last night. Got home at about 3 a.m., I think he said.

Brought more mail from Montgomery Ward Co., seed catalogues, letters, etc., but nothing from Dr. Holland, which was a disappointment as we are nearly out of money.

Got "Bailey's On Gardening".

Friday, January 28, 1910

Mr. Thorne was here and he said it was −10 the night before and −16 the night before that. The coldest he had noted at his place was −30

He brought over 6 dozen eggs. His hens have laid all winter. He said he got 50 cents per dozen in Vernal but he let us have them for 40 cents. We had a quite interesting talk with him and he stayed for some time. Took lunch with us.

The Thorne Ranch, located on the opposite side of the Green River and about two miles upstream, was the only other ranch as close as Neals'. Thornes were seen only on occasion, however, except during the winter when the river was frozen over and could be crossed on the ice.

Wednesday, February 2, 1910

Blustery after 4 a.m. or earlier. The wind would roar down the gulch and strike against the tent with hard puffs. Joe Arnold brought a load of hay; says the wind is blowing hard and the roads fill up with snow outside the gulch.

We all went to work but the go-devil broke down and we concluded we would have to fix it before we could proceed. Joe went to Jensen to get bolts, irons, and some provisions. At last I got a letter from Dr. Holland. He's deposited $600. I think that will just about pay up to date.

Friday, February 4, 1910
　　Mighty cold last night. Mostly clear during the day.
　　The work is progressing finely not withstanding the trouble breaking eveners, whiffletrees, etc. Have got abut 40 ft. of our trench dug. This evening had a nice smooth face of clean rock where we were excavating. Jim put in one shot and tore it up all along the face. The best shot he's used I think. The trench is about 20 ft. wide.

Tuesday, February 8, 1910
　　Mr. McHugh and Clarence went up on the hill to work. Pearl and Gawin went with me as far as Neal's. Mrs. Neal is sick.; has trouble with her heart.
　　Mr. Neal and I went up Brush Creek to the coal mine which is under the management of Mr. E. Burton. We wanted to find if there was an extra mine car that we could get for hauling dirt.
　　We then went to Mr. Bowen's and he said we could take his car. His son and Mr. Evans helped load it and we brought it home. I stopped at Mr. J. Murray's and got 2 dozen eggs, 4 lbs. butter and 225 lbs. of oats. Bought a bushel of potatoes of Mr. Neal.

Wednesday, February 9, 1910
　　Weather the same. Cold and fair.
　　After breakfast Clarence drove around to our east road and up to the trail landing with the car. We unloaded it on the platform then the other boys went down to Goose Island (Sandy I.) with team and got cottonwood for ties.
　　After lunch all went up the hill. Jim and I worked on the trail. Then Joe went with Clarence and they, after some difficulty, got track for the car up to the cut. Then Mr. Arnold came and I came down. The rest of the time was busy unloading supplies and getting things arranged. This is the first large lot of supplies since November.

Saturday, February 12, 1910
   Wrote to Dr. Holland telling about financial and other affairs. Sent for books which I very much want:
   Tanganyika Expedition
   Kellogg's   – American Insects
   Marshall's – Mushroom Book
   Hislip's   – Science and Future Life
   The boys are nearly ready to lay the first rails for the car track

Monday, February 14, 1910
   Started a new era this morning. We have a new repeating alarm clock set in the dining room and I hear it in the sleeping tent. I got up and built a fire. Pearl got up and got breakfast and we had breakfast by 7 o'clock. Got through by 7:30. That gave time to rest a little while and begin work on time. It gave us a chance to put in a good day's work and we did it and Pearl had time to do her work and do it easy.
   Heated water this morning to thaw out the barrel in the cave. It has been frozen for months. Worked on our new cut for car track.

Monday, February 21, 1910
   Sun warm for a time in the afternoon.
   Excavating is progressing nicely. They will soon be up to where we excavated with the horses.
   This evening we have been studying and talking garden. Made a rough plot of the ground, marking approximately where we would put vegetables and flowers.

Wednesday February 23, 1910
   At last there is a change in the weather. Last night when the sun went down it did not get cold. Froze but very slightly, if at all. This morning the air felt different. There was a faint freshness as of coming spring.

Mr. Arnold came at about noon and brought about 900 ft. of lumber, a washer and wringer for Pearl, and some other things.

Thursday, February 24, 1910

Decidedly warmer. The winter seems broken at last. It has been somewhat cloudy most of the day but when the sun shows it is felt.

Pearl used her new washer and wringer and I helped her some. Put out a big washing and not terribly tired as usual.

We have 15 hens and a rooster now. Had trouble to get them into the coop last night but they went in tonight. Yesterday we got an egg. It seems like home in earnest now. The rooster is as good as an alarm clock. It seems a little queer to hear one here.

With the long cold winter, which had taken its toll, finally broken and the advent of spring near at hand, my folks enthusiastically looked forward to the summer not only for the chance to uncover new dinosaur bones but to develop the camp and raise a garden. They were also soon to begin to dream of acquiring land for a ranch in the wilderness. Although their enthusiasm was to be somewhat dampened by a sudden visit from Dr. Holland, it would not be for long.

⌘ ⌘ ⌘

30

A DREAM COMES TRUE:
An Outstanding Discovery

Wednesday, March 2, 1910

There are lots of insects now, especially flies. Spent most of the time, regular hours, at the Dinosaurs though I did considerable work, writing, etc. down here. Worked from dusk in the morning until dusk at night. This evening I spaded some garden and put dirt on the ice house. I do enjoy working here. If we stay I intend to make this one of the coziest, prettiest places one ever saw.

Have been fixing up the part of the neck of the large dinosaur which we got down last fall. Have been trying to think of some better scheme to use with these large heavy blocks. The amount of plaster is appalling and one cannot get under to cement the block before taking it down. I think I have hit on the scheme. I am going to use woven wire and I am sure it will work.

Heard and saw wild geese.

For the benefit of those who have not visited Dinosaur National Monument, or are not familiar with this type of fossil collecting, I will give a brief description of the procedure of removing the bones from the sandstone ledge and preparing them for shipment to the Carnegie Museum.

Practically all the bones in this particular area were contained in a six-foot thick layer of very hard, course-grained, sandstone rock standing at an angle of about 60 degrees from the horizontal and dipping to the south. This made it possible to excavate a cut, or trench, on the south side of the "bone layer", as it was called, in the softer sandstones. By being careful and not getting too close to the bones, blasting powder could be used in this phase of the work. This excavation was done at first with picks and shovels, and teams and scrapers. Later, as it became necessary to go deeper, the excavation was all done by hand drilling, blasting and hand mucking into a regular mine car on a track.

After the bone layer was exposed by excavating the work was all done by hand, no powder being used. The cliff, or wall of sandstone, was now carefully worked into by slabbing off small blocks of rock and carefully watching for any indication of bone. This was done by drilling a line of holes some four to six inches deep and using plugs and feathers, which was the same method used in those days in stone quarries for breaking rock. The feathers were flat pieces of steel and were inserted in the holes in pairs, and the plug was a steel wedge which was driven between the feathers with a sledge hammer. By driving a little at a time on each plug a slab of rock was eventually broken off.

When bone was found the work was then all done by hammer and chisel. The rock was extremely tough and hard to break but the bones were in a wonderful state of preservation and the rock freed from the bone very readily. This was extremely tedious work. Sometimes it would take weeks just to work around one bone. Any pieces of bone broken off in this process were temporarily pasted back in place with flour paste. As soon as the bone was uncovered it was treated with shellac and alcohol to protect it from any possible deterioration by the elements.

After the bone was uncovered enough on one side to determine it's exact position it was covered with plaster. This was done by first placing tissue paper and shellac on the exposed bone, then covering it with strips of burlap dipped in a thin mixture of plaster of Paris and water. Woven wire and more plaster were used on the larger blocks for further reinforcement and protection. After the plaster had dried, the bone was then ready to take out in a block of rock. This again involved the laborious job of hand drilling and using the plugs and feathers. After the bone was freed in a block of rock it was completely plastered around for further protection and was then ready to be let down and placed in a heavy lumber crate with skids. The crates were skidded down the hill to await shipment to the Carnegie Museum. A large pelvis, which almost had to be taken out in one block, would weigh several tons. Each skeleton, each bone, each block was numbered and carefully charted in its original position.

Friday, March 4, 1910
Used wire netting for specimen of tail with 4 vertebrae. I am sure it will be a great benefit here, will save money, make the specimens more safe. Find it is going to take an immense amount of material - shellac, lumber, alcohol, burlap, and plaster to put up the specimens.

Sunday, March 6, 1910
Snow mostly gone except from sheltered places.
Went to prairie dog town but did not get any good pictures. Stopped and took pictures of the ice jam on the river. Worked here in the afternoon clearing up yard and getting things ready for our garden. I am tired tonight and not able to do as much as I would like.

Tuesday, March 8, 1910
Warm as usual. Windy this evening.
Went to work on the last find this afternoon. It is a little dinosaur probably laosaurus (editor's note – now known as Dryosaurus. SAB).

Friday, March 11, 1910 - Weather still the same; snow mostly gone.
I expect to go to Jensen tomorrow and to the sloughs beyond to hunt ducks with camera.

Sunday, March 13, 1910
Joe came home with me last night as we concluded that we would work Sunday to make up for yesterday.

This afternoon we went to work on our irrigation plant. We had talked of it a long time. Joe started to make a head gate; Clarence to nailing boards together for a trough and I went to digging a reservoir. Got water past the house at dark.

Thursday, March 17, 1910
We killed and dressed the pig this morning and began our summer room on the north of the house.
Clarence and Joe went to Jensen to help celebrate St. Patrick's Day this p.m.

Monday, March 21, 1910
In the forenoon uncovered the large dinosaur and began shellacking the one just found and getting ready to plaster and make it more secure. To get them out in blocks so we can handle the bones is the worst problem I have had and I am anxious to get it done.

Wednesday, March 23, 1910
Windy last night. I didn't know but it would wreck the camp. It came in gusts here.
This afternoon plastered spines of No. 1. I have concluded to try to get the ilium and sacrum together. It will make a block 7' x 7', say 5 cu. yards at 2 tons per cu. yd. would be ten tons. Perhaps this is a high estimate. If I could get that boxed and down the hill I would think I could do the rest.

Wednesday, March 30, 1910
I think we have solved another important problem that has bothered me. There have been two hard puzzles in connection with this work: to get the blocks enclosed so they would hold together and to box them securely. One problem I have solved with woven wire and plaster and the other by not boxing.

We made a bottom of 2 x 4 cross pieces and boards lengthwise and are crating things. I think the crates will be better in every way than the boxes and they will take only perhaps one third the amount of lumber. I think by shellacking the plaster the rain will not hurt them.

Tuesday, April 5, 1910 - Weather comfortable
Pearl not well and I helped her so got up to Dinosaurs late. Worked around the portions of No. 22 which have not been worked out. The boys worked at the cut.

Saturday, April 9, 1910 - Another nice day
Getting ready to shoot behind No. 1. Plastered ends of tail bones and a specimen a little distance away which I had thought might be tail. Took up a chunk and sure enough it was a nice tail bone. Picked up another chunk and there was the next in exact position. I waved my hat and hollered. The boys came. It looked as if after all we were to get the tail of No. 1. I had half given that up. I dug a little for the third. Guessed it ended there. Dug a little more. Hurrah! It was there in place. A little space between, evidently for cartilage. I said, "There never was such a specimen!" It does look as if we are to get an almost complete skeleton.

Tuesday, April 12, 1910 - Mostly cloudy and comfortable.
Received a letter from Dr. Holland yesterday. He is ill and his temperature was 102 1/2° when he dictated the letter. I do hope he is better.

Monday, April 18, 1910
Joe came this morning but was sick and not able to work. He brought a telegram from Dr. Holland stating that he would start the evening of the 15th. If so it may be possible that he is in Mack tonight and may get into Vernal tomorrow night. I may drive to Vernal this evening or Tues. morning.

Tuesday, April 19, 1910

Weather fine. Cleaning up the cave and around the diggings in forenoon.

When I came down to lunch or when I was on the brow of the hill I saw a buggy and someone waved his hat at me and said, "Hello old man". As soon as he spoke I knew it was Dr. Holland. Was glad indeed to see him.

After lunch we went up to the Dinosaurs. He seemed struck almost dumb by the magnitude of the work, the labor necessary to get the things out. I felt disappointed as I felt so sure he would be greatly pleased. He did not say very much, however.

Wednesday, April 20, 1910

Dr. Holland spent the greater part of the day at the diggings. He does not feel so confident as I do of complete skeletons and he feels that the bones are much disturbed as in other places. He says it is the best find since the discovery of the one on Sheep Creek. I believe it is better than he thinks it is but he is not going much on the unseen. I have worked here enough to have much confidence. I certainly believe we will find at least one or two skulls and that would put this place ahead of any other. I haven't lost faith but wish he felt as I do. He says I can go into the Wasatch when I get through here.

Thursday, April 21, 1910 - Weather fine

Working at the Dinosaurs. Dr. Holland with us part of the time. Worked out ilium of the dinosaur smaller than brontosaurus and found just a little distance from the acetabulum, the head of a large femur which is 21 or 22 inches across the head, the longest diameter. I thought it was the long-lost femur of No. 1. Have felt much disheartened a good deal of the time since Dr. Holland came, yet he is interested.

Sunday, April 24, 1910

Irrigated garden in the forenoon. Pearl, Gawin and I went and took a walk. I took pictures of flowers. Tried to find bird nests but didn't succeed. There are many kinds of birds in the gulch now.

Got ready after lunch to go to Vernal to take Dr. Holland. I think it was between 2:30 and 4 p.m. when we started.

Dr. Holland stopped on the road to catch insects. Got into Vernal about dark,

Took supper at Heidelberg Restaurant. Roomed at Alwilda Hotel.

Monday, April 25, 1910

Got up early so Dr. Holland could take the stage. His visit has brought back some of the old feeling toward him. I think he is unjust and unappreciative. We expected approbation, for we knew we had been faithful, but got little but censure. If I had another position and these things were not so interesting I would quit and see if he could do any better. There is no use in a man being so unappreciative when it is as much for his honor and that is what he wants. I do as much as any man on the force and yet he is always digging around.

Got home at about 12 midnight. Bought a wagon $109.50 and engaged a harness. Did not buy a team but think perhaps Mr. Logan has just what I want.

Sunday, May 1, 1910

Rained last night and a part of the day.

We arose late and it was about 9 o'clock when we ate breakfast. It was a good day to put in garden. Jim said he would help me so Clarence grubbed, Jim spaded the ground and we worked hard until nearly dark.

Put in 2 kinds of water melons, 2 kinds of musk melons, 2 kinds of corn, some potatoes, 2 rows of peas across the garden, spinach, turnips, rutabagas, 3 kinds of cabbage, 3 kinds of cantaloupes, okra, kohlrabi, 2 kinds of squash, cucumbers, beans onions. Did not have room for all I want to plant.

Tuesday, May 3 1910
Pearl very sick. I had to stay with her, take care of Gawin and do the housework. Found it a very tiresome task. Jim went to Jensen and tried to telephone the doctor but could not reach him.

Wednesday, May 4, 1910
Went up to the quarry a little while but dared not stay.
Jim took team and went to Vernal. Took Gawin to Mr. Neal's as it was almost impossible to keep him away from his mother. They seemed glad to have him come.

Thursday, May 5, 1910
Got along a little better. Jim did not come last night. Did not get here until dark tonight. Brought some medicine prescribed by Dr. Martin. Went down to Neals' to see Gawin. He is getting along alright.

Wednesday, May 11, 1910 - As usual, hot in p.m.
Pearl feeling no better. Clarence went and took her to Mr. Neal's today after lunch.
Have been at work at ilium and vertebrae of No. 1. Have a nice little job with them but they are coming out in fine shape.
Our garden is growing nicely. I have seen none as good. Had radishes for breakfast. Peas in bloom and are a foot or so high.
The boys brought home a big fish tonight. It probably will weigh 20 lbs. or more.

Wednesday, May 18, 1910
Went to see Pearl this evening. I think that on the whole she is better though she suffers a good deal at times.

Have put in a good days work today. Struck the tail of the new dinosaur about 11 feet back of the pelvis. It is smaller than brontosaurus. I don't know what it is unless it is mososaurus (editor's note: I think he is writing Morosaurus – which is now Camarasaurus, a common critter in the quarry, SAB). It isn't like diplodocus.

Thursday, May 19, 1910
We are changing the complexion of things rapidly now.
Have numbered the new dinosaur (Morosaurus?) No. 40. Have been working along the line of that dinosaur and on both sides of the pelvis of No. 1. I hope we can take this pelvis down in a few days. Though much of the work is hard the things are coming out pretty fast.
We have some failures. Our ice is melting and we may not be able to save any. As we expected the chickens are getting after the garden. Have eaten nearly all the young radishes down to the ground. Intend to shut them up.
Am busy from morn' til bed time.

Tuesday, May 24, 1910 - A nice day, showery around us, thunder
Mr. Neal sent word by Joe Ainge that he had engaged Thorne's boat and would go fishing in the morning. I concluded that we would go for a half day as we are entirely out of good meat of any kind. Have nothing but a little salt meat and that not perfectly sweet. I wanted to see fish-seining for once, get pictures. We made fine hauls and got between 50 and 60 fish.

Wednesday, May 25, 1910
Digging is getting more interesting; hate to quit at noon and night. Worked on cervicals of the large beast. The cervical ribs are about 18 inches long. It surely is a huge neck. I doubt yet that it belongs to No. 1, though I have been more inclined that way lately. Have caudals of No. 40 numbered to

25, I believe. The stratum is full of bones. Bones are going diagonally through cliff. It looks like a delta deposit and the current to the east.

Sowed red flowers next to prize beans. Daisies, asters, coxcombs, carnations. Radishes that the hens ate off are coming in very nicely.

Saturday, May 28, 1910
Concluded I would go to Vernal and see if I could buy a team as Mr. Goodrich wanted his team returned by the end of this month. Got into Vernal before sundown, stopped on the way trying to find horses. Had no success and was having no success in Vernal. Charley Neal told me of a mule team.

Sunday, May 29, 1910
Went to Charley Neal's, got a riding horse and went up to Snell Johnson's. I liked the mules and made out a check for $400, the price asked.

I was very busy all morning. Johnson sent the mules down and I got collars fitted.

Wednesday, June 1, 1910
Partly cloudy but awfully hot when the sun came out.

Am working at and around pelvis of No. 1. Do not think I have found the skull of No. 25, will probably be the neck of No. 1.

All the boys working today. Others working on No. 40, which is a magnificent specimen. Never saw the like.

The puzzle concerning the neck of the brontosaurus (No. 1) continued. It was thought to be another dinosaur for some time and was given another number (No. 25). When it was found that the anterior portion was under the pelvis of No. 1 there were high hopes, for a time, that the skull would be found in place. This was not the case, however. The skull was never found.

Thursday, June 2, 1910
　　As usual hot and windy in the afternoon.
　　Finding new bones nearly every day we work. In morning we went up to spring and made a reservoir. It is getting so we can hardly get water in the middle of the day. With the reservoir I think we will have little trouble in irrigating the garden. Some of it is pretty dry now.
　　The garden is quite useful now. Had radishes May 1, onion sets in April, beet greens June 1, peas June 2. Pearl had a little lettuce yesterday.
　　I can't help thinking about our last dinosaur.

Sunday, June 5, 1910
　　Got up early got ready and Jim and I went down to Jensen slough or the little one below old Jensen ferry. Pearl and Gawin went down to Mr. Neal's and stayed all day. Had the greatest sport I ever had with the camera taking pictures of blue herons wading in the marsh among the reeds, finding nests of mud hens and blackbirds.
　　Went to Jensen on our return and I bought some things. Went to Mr. Snow's and got Miss Snow, who is helping Pearl.

Tuesday, June 21, 1910 - Hot! Hot! Hot!

Tuesday June 28, 1910
　　Had a little shower in the morning and we had one this evening after dark. The water came rushing down the gulch. I suppose our dams are all gone. Will have to fix them for we have to have water.
　　Right after breakfast it began to look like rain and sprinkled some so we hitched the mules to the go-devil, fixed the crossing and hauled bones. Came near having an expensive accident. A stone was caught by the whiffletrees and rolled under the go-devil when we had on a heavy box. It rolled the box off in front and almost on the mules. I expected to see their legs broken.

Friday, July 1, 1910

Jim, Joe and I working; plastering, digging, making boxes.

In the evening Mrs. Douglass had a bad spell with her heart or whatever it is. It was the longest spell of the kind she has had I think. I was pretty scared. Della (Della Snow the hired girl) and I worked with her and finally she got better. She slept soundly all night but was weak and not well this morning.

Saturday, July 2, 1910 - Hot.

Mrs. Douglass weak and feeling pretty badly. Sent Joe to Jensen to telephone the doctor and get some things. Jim and I worked at the Dinosaur in the forenoon. After dinner we went up to fix the reservoir which had been washed out by high water and heavy shower. The spring is good as ever.

At last we have settled the problem of the pelvis of No. 1. Have got the left ilium off. It came off in good shape. It has removed a burden from my mind.

Sunday, July 3, 1910

Arose late in the morning. All wanted to rest, I guess. I know I did.

Mr. and Mrs. Murray came during the day. Took lunch with us. Pearl had another of her sinking spells but not as bad as the last one.

Thursday, July 7, 1910 - Hot.

We were going up to the Dinosaurs after supper but Dr. Martin came to see Pearl when we were eating our 5 o'clock dinner. The doctor thinks Pearl will soon be better; is going to send some medicine. There is no organic trouble with Pearl's heart. It is the nerves which control it.

This was the beginning of a heart condition from which my mother never fully recovered, and which caused her much trouble throughout the rest of her life. We always felt that the cold winter of 1909 and 1910 contributed considerably to it.

According to my mother's account the winter was much worse than would be gathered from my father's diaries. She related that there were days and days on end when they were never really warm in the hastily set up tents. Of course her greatest concern was my welfare and she probably sacrificed her own health for mine. She related how she would get the fire going full blast in the cook house, which was nothing more than a boarded up tent, and hastily bathe me in a small tub on the kitchen table.

The cold of the winter, however, was little less extreme than the heat of the summer and the two, together with the hard work of maintaining the camp and cooking for the men, were probably the factors that affected her health. Apparently the elements had no injurious effects on my physical well-being whatever.

In spite of all the hardships Mother never intimated that she would have had it any other way. It was her desire to be with my father and his work no matter what the conditions were. I never heard either of them say anything that would suggest that they would have rather spent the winter in their nice home in Pittsburgh. Through the cold, the heat and illness the work continued at the quarry with increasing enthusiasm. New problems were solved and new discoveries were made almost every day.

>Monday, July 11, 1910
>
>Clarence returned. His father brought him over with the horse and buggy. A lot of people came from Naples, picnicked by the Lone Tree and came up to the Dinosaurs. Attorney Hansen was there and several from Vernal. It was about the largest crowd that has been there this year.
>
>Thursday, July 14, 1910 - Mighty hot
>
>This has been one of our big days. We got the pelvis of No. 1 all out except the upper part of the right pubis and most of it is down the hill. Got three big boxes down today. This morning the sacrum was up on timbers. We hitched

the mules to it and slid it down, then put the go-devil on top and the chain around and turned it over with the mules bringing the specimen on top in place on the go-devil. Had a time getting it up the little pitch at the bottom of the hill but worked it by blocking up and putting rollers under. It seems that now we will have no more such heavy blocks for a long time.

Saturday, July 16, 1910
They say it was 104 degrees in the shade in Jensen and 118 in Vernal.
It took nearly all day at the quarry wrapping, numbering, labeling and plastering specimens.

Monday, July 18, 1910
About noon Mr. Young and Prof. Elmer Riggs (paleontologist from the Field Museum in Chicago, who was collecting in the same area as my father in Montana several years previously, GED) came from Vernal. His camp is in the cabin where Frank and I stayed two years ago. He says they have been having pretty good success the last two weeks. Have found a skull of Uintatherium and titanotherium material.
He expressed little or no surprise at what we think are great dinosaurs. In fact, made no words of commendation, though nearly half a dinosaur was practically in view, and a good part of another. He seemed a little queer to me.

Friday July 22, 1910
Found that all, or nearly all, of the ribs of No. 40 on the left side, are in place attached with vertebrae. A wonderful specimen.

Saturday, July 23, 1910 - Terribly hot
Joe went down to Jensen in the afternoon to get some plaster but Mr. Snow brought it so his trip was for nothing. Got the mail. Ida is failing all the time and I expect any letter

I get to hear that she is gone. It will be a relief to hear that she is past her suffering. No words can express the damnation of this life, some phases of it. Anyone who looks only at this life and isn't pessimistic is a fool or hasn't much memory. Yet I admit that it may be that life continues, though it sometimes seems plain it does not.

Monday, July 25, 1910 - Hot.
Boys all away today celebrating the Mormon holiday so I have had all the work to do in camp. Have been redrawing dinosaur No 1. This will be quite a job. Will probably have to do most of it at the quarry.

Have one sunflower that I cannot hang my hat on from the ground.

Monday, August 8, 1910 - Very hot again
Flies and heat so bad it is hard to rest in the middle of the day.

Did not go up to the quarry today. Worked on diagram of the bones; cleaned and straightened out the store tent that was in a chaotic condition, and helped Pearl some as she washed this forenoon.

Yesterday we had a dinner from our garden and it was fine; summer squash, mashed potatoes, string beans, green corn, cucumbers, beets. Today we had cabbage, okra, onions in soup, turnips, potatoes and radishes.

There are almost swarms of birds in the gulch of a number of different kinds. Many of them are feeding on buffalo berries which are now ripe.

Wednesday, August 17, 1910
A pretty hot afternoon but with a breeze from the west.
This is the first anniversary of the discovery of the fossil quarry which is undoubtedly destined to be famous. I intended to, in some way, celebrate the day but forgot until afternoon.

There has been a big change since a year ago. We have had a big task and it is not yet by any means accomplished. We have out the larger part of No. 1, if the neck (25) belongs with it, but we have not found the skull and forelimbs and feet.

We did not begin work until September 2, that is with a force of men, so we have two weeks yet of a year's work to do. I think the next 2 weeks will be eventful. We are getting the cut where we can get dirt away and see what we have in the western part of the diggings.

Friday, September 2, 1910

It was one year ago today that we began work with force, Clarence Neilson, Jim McHugh, Joe Ainge and myself. We have done a lot of work. Our expense has been between $5000 and $6000. When I began I estimated the cost of getting out the one dinosaur, No. 1, at $2000 to $5000. We have gotten out the greater part of No. 1 but not all by any means. We also have the greater part of the tail of No. 40 besides several specimens smaller which would be called skeletons and dozens of other individuals represented by parts. We have run a cut over 100 ft. long in hard rock. In some places it is 35 or 40 ft. high on one side. We have established our camp, etc.

Unfortunately we have not yet found a skull. I hoped to do so before this and have had many visions of it.

Thursday, September 8, 1910

Pearl, Gawin and I went to Jensen this morning to get several things, mail, letters, etc. Got a letter from Jay telling more about Ida's death. They are talking about moving away. That would leave Nettie and Father alone in their sorrow. I must, of course, see that Nettie and Father are made as comfortable as possible.

Got a letter from Dr. Holland which made me angry again and I would leave if I were able but on account of others I must hold my job.

Mr. Kay (older brother of Dr. J. Leroy Kay) went to work today.

Monday, September 12, 1910
Went to Jensen in the evening, Pearl, Gawin and I. Did not get any thing we went for and Pearl and I both came back sick. I haven't had a worse spell for years.

Sent a letter to Dr. Holland telling him what I thought. I felt it was positively necessary as I looked for trouble to come to a head when he comes. Maybe I ought to be different but can't stand it with him. Unless he is different we probably will all leave, but of course we have a lot to lose if we do.

Saturday, September 24, 1910
Am getting along very well with the dinosaurs. It seems now that No. 1 is turning and running down. The large femur may belong to No. 25. instead of No. 1. We have a rib that is 7 ft. 4 in. long.

Have got down a foot or two farther on No. 40. It looks as if it is rising out of the ground and almost startles a fellow when he looks at it.

Sunday, September 25, 1910
At home most of the day. About 4 p.m. all who were here went over to McHugh's Gulch. I wanted Pearl to see it and I wanted to see what the prospects were for taking water right and land for a little nook for a dwelling place. One could make reservoirs, I believe, and make a beautiful and romantic little spot of it.

Thursday, September 29, 1910
Boys went to Jensen in the evening; got mail, butter etc. Got a letter from Dr. Holland. He appeared to feel much grieved by the plain letter which I wrote him. It had got where something had to be done. I couldn't help thinking about some things and feeling bitter. I think his letter was

written in a very good spirit considering. We were pretty glad that I was not fired.

Got prints and enlargements from Eastman Kodak Co. The prints were excellent. The enlargements made good pictures, part of them. They are expensive.

Wednesday, October 5, 1910 - Nights cool, build a fire to dress by every morning.

Much of the garden is still green; still have melons. I picked one of our Hubbard squashes the other day and it was dry and fine. I was surprised that it is so good; still have cabbages.

Have been doing great execution at the quarry. Got out femur of brontosaurus No. 1, that of diplococcus No. 60, the rest of the scapula of 25 and parts of 3 ribs. It has changed the face of the quarry considerably. Don't know yet whether neck 25 belongs with no. 1.

Thursday, October 20, 1910

This was our wedding anniversary; have been married 5 years. Have never had a quarrel and we are as much or more to each other than ever. On the whole we enjoyed the day and talking over old times. I gave Pearl a silver butter knife and she gave me some collar buttons.

Sunday, October 23, 1910

We got up late and came to dinner late. When we came down we saw Dr. Holland nearly in the same place as he was when I saw him first when he came before. After dinner we went up to inspect the quarry.

Thursday, October 27, 1910

Dr. Holland and I took the mules and went down to Jensen to see about getting freight teams to haul out boxes. Went to Mr. Wall's and engaged 3 or 4 teams there. Went to

Mr. Rasmussen's and engaged 3 or 4 teams there and one of Bert Case.

Andrew Carnegie and the members of the staff of the Carnegie Museum were now anxious to have the bones of the brontosaurus in the laboratory of the museum so that the tedious work of preparing the skeleton for mounting might be undertaken. Thus preparations were made for the first shipment from the quarry during the early fall of 1910. Dr. Holland returned to Utah to help supervise the task of getting the bones to the railroad. Transporting the dinosaur out of the Uinta Basin was no small undertaking in those days. First the heavy boxes, or crates, had to be loaded on wagons, drawn by two or three teams of horses, and hauled some sixty miles over rough and hazardous roads to the Uinta Railroad terminus at Dragon. Here they would be transferred from the wagons to box cars. They would then be transported sixty miles by the Uinta Railroad, which was narrow gauge, to Mack, Colorado where they would be reloaded on the standard gauge cars of the Denver & Rio Grande Railroad and shipped to Pittsburgh. As will be seen, the task of loading and reloading was done principally by my father and his crew.

> Monday, October 31, 1910
> The teams came pretty early but we were prepared for them. Had an awful time getting the heavy specimen of the carnivorous dinosaur into the wagon box. The bottom broke out and the boys wired it up. The large specimen of cervical did not bother much until it was in the wagon then it nearly tipped over. Nearly all the horses balked and the result was the fellows (dinosaurs?) did not all get loaded. Dr. Holland feeling very badly.

> Tuesday, November 1, 1910
> Dr. Holland went away this morning.
> We arose earlier than usual. At last between 10 and 11 a.m. the fellows got out of the gulch.

Thursday, November 3, 1910
   Started for Vernal

Drove the mules to Jensen and left them in Joe's brother's field so we could get them when we returned. Went to Vernal with Mr. Snow who carries the mail. Got a room at the Alwilda.

Dr. Wilson, the veterinarian, wanted me for a lecture. He told Mr. Young and the latter was a little "riled" at first. Mr. Young has for a year or more wanted me to lecture on geology but they had the evenings provided for. He wanted me to go over to the hotel and see the State Supt. of Public Instruction and Professor Hall. I was glad to meet them. They soon arranged for me to follow Prof. Larsen, the lecturer of the evening. I went to the hotel and told Pearl. She insisted that I must have some new trousers. We went to the Co-op and the result was I got a new suit of clothes, a fine shirt, collars, ties, new shoes etc. So I got fixed up pretty well after coming out of the woods.

I did not feel much of that palpitation which I usually feel before a public talk. I had a large audience.

Friday, November 4, 1910

Started in the morning from Vernal in the auto which carries passengers and mail. It was a pleasant change to ride in the automobile but we had about enough by the time we arrived in Dragon.

Sunday, November 6, 1910 - We are at the Mack Hotel this evening.

We finished loading cars and got things fixed up in pretty good shape. Had one box car and one flat car. If we had gotten all the things out which we expected to we would have, as I predicted, a Uinta trainload. As it was they pulled our freight and one car of Gilsonite.

Tuesday, November 8, 1910 - At Mack

Found there were Pennsylvania cars there. Ours was 55088. The transfer boss with his crew could not be ready for some time. We saw our mistake after we got to loading. We ought to have taken one of the crews and done the things to suit ourselves. The Greeks were very poor hands at handling that kind of freight and it got pretty badly shaken up.

Wednesday, November 9, 1910

Started on the morning train for Dragon. We arrived in Dragon a little late. Had to wait some time for the auto. We rode with express and mail. Our auto broke down at White River. At Kennedy we had a flat tire and on top of Dead Man's Bench punctured a tire and had to make a fire of shad scale and put on a new tire.

Friday, November 11, 1910

Came home. Were glad to get back to the gulch.

Monday, November 14, 1910

I began to investigate the teeth we found several weeks ago. Worked under the shoulder blade and found thin bone. Traced it up and it seemed like the lower border of the upper jaw of diplodocus. Hoped it was, of course.

Tuesday, November 15, 1910 - SKULLS!

Dreamed the greater part of the night of working on diplodocus skull. . .uncovering the bones of the back portion time after time until morning. I would awake and get up and then go to bed and dream it over and over.

I was anxious to get up there and investigate. I uncovered or traced other, right side of jaw, to back then went down and found the occipital condyle so before noon was sure of a skull, at least lower portion. Was jubilant!

Had set Steve Murray to work around the remainder of the little dinosaur in the cut. I went down in the morning and looked it over -- no bones appeared. Joe went down in the afternoon, saw bone in one place. I went down later, raised the plaster up we had put on, took out a piece of fresh rock, and behold a SKULL.

Wednesday, November 16, 1910

Getting things in shape ready to pull out. I wrote to Dr. Holland about the skulls. I am so glad to get these as this "Caps the climax", even if I did not find them until the 11th hour. This leaves little to be desired. Of course not all the good things are out but many are in sight.

The activities at the quarry were soon to end and would not be resumed until the following spring. It was necessary for my father to return to the museum with maps, notes and other information for the preparators so they might embark upon the gigantic task of mounting one of the few complete skeletons of a dinosaur in existence.

After my folks return in the spring of 1911 the activities at "Dinosaur" were to be pursued with great enthusiasm, not only at the quarry where undreamed of discoveries were to be made, but to fulfill a personal dream of acquiring land and building a home in the wilderness. The West in general, and especially the barren but colorful country surrounding my father's dinosaur discovery, were soon to completely captivate him.

31

CAPTURED BY THE WILDERNESS:
Dreams of a Ranch

Before the first departure from Utah since the dinosaur discovery, all exposed bones were covered with plaster and the two recently-discovered skulls were removed. In late November the diplodocus skull was being worked around.

> Tuesday, November 22, 1910
> We got up in good season this morning and got in a good days work. I worked around the skull of diplodocus so I think it will not take long to get it loose.
> After dinner we plowed the garden and such deep rich sandy soil. It does one's heart good to think what we can raise on it... am anxious to plant a garden... will have a much bigger garden than this year.

On November 25th, after hurried preparations, we departed. We left the Uinta Basin via stage from Vernal and went westerly to Price, Utah instead of taking the southerly route over the Uinta Railroad to Mack, Colorado. At Price we boarded the train for Montana. After spending a few days in Montana with my grandmother, who was in poor health, it was decided that my mother and I would remain there and care for her instead of going to Pittsburgh with my father as was originally planned. The parting was not pleasant as indicated in the diary.

> Wednesday, December 7, 1910
> Took the train at about two o'clock for Whitehall. Parting was hard for Pearl and me. At the last moment nature got the better of me. I cried like a child. Pearl felt so bad and it

was so hard to think of little Gawin without a papa. We felt though that our duty was clear. I could not say anything to Mother.

Friday, December 9, 1910
Arrived in St. Paul on time. Came on the nine something R. I. train to Medford.
I had a message from Dr. Holland, an insulting thing wanting me to come as soon as possible and help straighten up matters. He said there were not as many boxes as I had on my list and I had greatly overdrawn my account.

Tuesday, December 27, 1910
Was detained in Medford ten days after I should have gone to Pittsburgh on account of the telegram received from Dr. H. saying I had greatly overdrawn my accounts. I looked over my accounts, added amount of checks, subtracted from amount of deposits since the last balance and according to my accounts have not overdrawn, that is including my personal deposits.

Saturday, December 31, 1910
Getting things straightened up so as to get away. Traded cows with Wallace Hayes as his was fit for beef and could get rid of her at once. Sold her to George Lee. She weighed 1090 lbs. and brought 3 1/2 cents per pound.

My grandfather, Fernando and Aunt Nettie (Earl's sister) accompanied him to Pittsburgh. They lived in his home in Pittsburgh while he was getting the brontosaurus specimen in shape so the laboratory men could go ahead with the preparations for mounting. The entries in the 1911 diary are rather incomplete. In February he mentioned the work at the museum.

Saturday, February 4, 1911
At the Museum getting records of the quarry in shape. Copying sub numbers in my record book. Then I have three

sets of cards to prepare. One for contents of boxes, one for field numbers with the box in which they are packed and one with record history of sub field numbers. Dr. Holland gone for nearly a week.

Later in February Father received a letter from Andrew Carnegie.

> Thursday, February 23, 1911
> Received a letter from Mr. Carnegie of which I am proud.
>
> > Two East Ninety-first Street, New York
> > Feb. 22, 1911
> > Dear Mr. Douglass:
> > Many thanks for your valued letter. Glad one more Scotsman has made good.
> > What a wonder that Monster is to be. Mrs. Carnegie and I expect to visit Pittsburgh in April or May and hope to meet you and your wife and boy, he was right about Holland sure.
> > Yours ever,
> > Andrew Carnegie

On March 19, 1911 my father, his sister and father left Pittsburgh for Medford, Minnesota. The morning of March 23rd he took the train for Montana. He mentioned that he was glad to leave Pittsburgh and wanted to be with his little family. From Montana we proceeded to Utah, arriving at the dinosaur quarry the first part of April.

> Monday, April 10, 1911 - Rain this night and in the morning.
> Are getting work at camp well along. Got garden fence well along. We put floor in the summer dining room today.
> Mr. Schuler has moved up to his summer camp among the willows. Mr. Kay is making a summer house a little below. They have a trail over the sand dune hills and snake the things up with single mules. Mr. Kay does not expect his family until next Saturday.

Thursday, April 20, 1911

We are uncovering new bones all the time. We may be getting the bones of No 45. Have an ilium and a lot of vertebrae. It keeps me busy most of the time attending to fragments, keeping records, plastering, pasting, etc.

Began writing poetry this evening. I want to write something every day I can.

This seems a fitting place for my father's poem, "Hymn of the Wilderness." It well may have been written at this time, though I am not sure of the exact date. The poem reveals his deep love for nature and his basic philosophy. In nature are the secrets of the universe and through the study of nature is man's best way to some of the answers he is seeking; God, the source of life, man's destiny. The poem also shows how completely he was being absorbed by his wild surroundings.

Hymn of the Wilderness
By Earl Douglass
I wander midst the olden hills,
Mid rude rocks carved by tempest's rage
In days unknown. A mystery fills
The primal scene, sublime with age —
Sublime, for Nature reigns alone
Upon her everlasting throne.

Here altars rise on every hand
That mystic reverence inspires,
And ancient, silent temples stand;
And though I hear no chanting choirs
From near and far there seems to roll
A mystic music to my soul.

Here, freed from man's contending thought,
That makes a din of hate and strife,
I find what long in vain I sought,

A nearness to the source of life;
Where scenes are fresh and thoughts are free
My deeper self returns to me.

Here are no books, no written word
That man ignores or blindly heeds.
No sound of human voice is heard,
No narrative of human deeds.
Yet to the one who list's are told
Strange stories of the days of old.

In this wild land of sagey plains
And sculptured, rocky-templed heights,
Where bigotry has left no stains
And tyranny no scourging blights,
Should not the poet love to roam
And freedom's lovers find a home?

With songs and hymns these rocks ne'er rang;
These hills the prophets never trod,
Nor sages e'er these wonders sang
Nor seers have heard the "voice of God"
Yet here the Spirit dwells, divine,
That spake with men in Palestine.

If prophet, genius, sage and seer
Would flash deep truths in human speech,
Would read the revelations here,
The Spirit's inspiration teach,
Their words would live through tempest shocks,
Their thoughts survive these crumbling rocks.

Father's crew now consisted of four men; three had been with him the previous year: (Clarence Neilson, Joe Ainge and Jay Kay), and a new man, Mr. Schuler. A diary entry in May gives the first real indication of a new era in our lives, the idea of acquiring land in the area, developing a ranch and building a permanent home. The almost undreamed of discoveries at the quarry, which indicated it should continue for some time, success with a garden the year before, and the fact that there was free land for the taking, all combined to make my folks eager to fulfil their dream of carving a ranch and home out of the wilderness. This was later to cause them much pleasure but also a great deal of pain and disappointment.

My father had discovered what he named McHugh Gulch the year before (later renamed Orchid Draw after a rare and beautiful species of orchid he found growing there). This little oasis was about a mile and a half to the west of the quarry. It is so well concealed among the many-colored sandstone cliffs that one little suspects its presence until practically stumbling into it. There was the upper flat and the lower flat, comprising some ten or fifteen acres of tillable land with a canyon between the two containing a nice spring of pure, cold water. Birch, cottonwood and box elder trees grew around the spring and along the gulch. There was a nice little grassy meadow below the spring. This was all in sharp contrast to the surrounding dry, barren hills and cliffs where only a few scattered juniper grew.

> Sunday, May 21, 1911
>
> A most pleasant day and long to be remembered. I did the chores and turned down the water from our new reservoir. It is a dandy. I irrigated nearly all the garden except the upper part. It's wonderful what a little spring will do.
>
> Pearl did the housework and got the lunch ready. I got the mules ready and we went over to McHugh Gulch. Drove up as far as we could and hitched the mules to the wagon. We took a quilt, bucket with lunch, and started up the gulch looking for a place to build a road. I had a grub hoe and fixed the path in places. It is a steep grade and in one place especially steep but I think with my crew of men I could get a road up in about 3 days. This would cost about 30 or 35 dollars besides my own work. Would have to have about $10 worth of lumber.

Made up my mind where the dam for the reservoir ought to be. It would have to be about 60 feet long but by excavating it would hold a lot of water. I think my crew could build it in 2 or 3 days. I think I could put 4 or 5 acres under cultivation. I think with a little water I could raise peaches on the upper flat above the spring. Dreamed a lot about it today and am now more impressed with it than before.

Wednesday, May 31, 1911
Pearl, Gawin and I went to Jensen and Alf Ainge's to get some wild hay for packing.
I telephoned the land office and found I could not take homestead in two parts at once so our plans were frustrated again.

The work at the quarry continued to progress with new bones being found almost daily. Mention is made of Mr. Jay Kay working around the bones of a dinosaur that did not appear to be either brontosaurus or diplodocus. Clarence was working around stegosaurus, No. 39.; No. 40 continued to be a marvelous specimen with the ribs in place on both sides.

On June 17th my father purchased a buggy for $165.00. He paid $25.00 down and $20.00 per month. He said that Pearl wanted the buggy and he felt he owed it to her. This was the white-topped buggy that appears in many pictures with the white mules. It was used for many years. The mules (Bill and Joe) pure white and well-matched, were the ones purchased from Snell Johnson May 29, 1910. They were an important part of the quarry all through the early years and became almost as well-known as my father. They were a good, strong team, but were so "tough-mouthed" and learned so many tricks, from having many different drivers, that they became almost completely unpredictable and uncontrollable. If a new driver was handling them and let down his guard for a moment they would take over and would soon be doing part of the driving themselves.

Every type of severe bit was tried but little was accomplished. Their mouths were as tough as leather. They soon became experts at opening gates, getting through fences and taking off bridles and halters. However, they were gentle as kittens with one exception. Bill, when being ridden, could not resist the pleasure of bucking a person off once in awhile, especially if the rider's hands were being used for anything except strictly taking care of the bridal reins.

One such memorable occasion happened when my father and I were returning from the gilsonite mines one summer with a wagon load of old burlap sacks to be used for plastering the bones. The wagon had a big umbrella attached, which was raised and lowered for shade as needed. It had been in use all summer. As we were plodding along the rough, dirt road, my father asked me to put up the umbrella to keep out the hot summer sun. When I raised it the mules suddenly bolted and started to run. I was holding onto the umbrella and having trouble staying on the wagon. My father had his feet braced and was pulling on the lines with all his strength. The team ran straight down the road at a terrific speed but made no attempt to leave it. This went on for some time. Soon we were approaching a place in the road that would surely wreck the wagon and throw us off and when I looked at my father his face was almost black from pulling on the lines. He gasped, "Let down the umbrella if you can." Struggling desperately, I did. Bill and Joe stopped just short of what might have been a tragedy and started walking down the road as if nothing had happened. Later we raised and lowered the umbrella several times without causing the slightest disturbance.

As the diary continues in July 1911 the flowers and garden were doing fine. We were eating young onions, potatoes, cucumbers, and green beans.

Mr. O. A. Peterson arrived from the Carnegie Museum to assist with the preparation of the bones. During the latter part of July McHugh Gulch is again mentioned, also an eventful day at Jensen.

> Sunday, July 23, 1911
> We hitched to the buggy and Mr. Peterson, Mr. Schuler, Stella Dodd and our family went over to our upper 80. Took

dinner by the spring. The water there is almost perfect, cold and sweet. The water does not diminish. I wanted to go for two purposes, to see if I could dig to water on the upper end of the 80 and to see if the ravines had reservoir sites. I got to very wet sand but did not get to water.

Monday, July 24, 1911
  A great day at Jensen. Never before, I am sure, was such a crowd gathered there. The new bridge was dedicated. It was not much fun but I was glad I went. I learned some things and we are now personally interested in the development of the country.
  It seems now that the Uinta region is beginning to shape up. Its possibilities are great. It only needs well-directed capital. I saw a good many I knew and many men I did not know.

This was the first bridge to span the Green River in the Uinta Basin, although the country had been settled some thirty years or more. All previous crossings of the river were made either by ferry boat or by fording.

A few entries relating to the garden (which was excellent), hauling down specimens and building a landing to accommodate them concludes the 1911 diary.

Dreams of trying to develop a ranch and build a permanent home started to become a reality for my father and mother in 1912. They finally filed on the eighty acres in McHugh Gulch under the Desert Act and the flat land to the south and west of the quarry, between the cliffs and the Green River to the south, was filed on as a homestead. This latter piece of land comprised over 300 acres.

Fired with enthusiasm and pride of ownership, Mother and Father began to build our ranch home: a two-room log house constructed on the flat at the mouth of the first gulch, west of the gulch where the present road now enters. The gulch was named Ainge Gulch after Joe Ainge. It had a spring some distance up the gulch from the house.

The location for the house was chosen primarily on my account. It should have been placed on the banks of the Green River by the Lone Tree, a huge cottonwood immediately to the south that had been a landmark for many years. The tree provided shade rather than being in the hot, barren ground at the foot of the almost vertical cliffs, which reflected the hot summer sun with more than its usual strength and water was close by. However, because of the fear of my falling in the river and drowning, the house was placed as far from the river as possible.

The Lone Tree was a thing of beauty and comfort during all the early years of excavation at the quarry, for no other trees near its size could be seen on the north side of the river for several miles. It was actually three huge trees of almost equal size, which grew practically as one. They were each some five or six feet in diameter and were limbless for about thirty feet from the ground. They then branched out symmetrically and formed what resembled a huge umbrella with its branches almost touching the ground. When looking at the tree from either the east or west the three trunks, being in line, appeared as one, but from the north or south three separate trunks could be seen. It is thought that Major Powell camped near here on one of his expeditions down the Green and Colorado Rivers.

At this point my father started keeping a special series of journals. One entitled "Diary 1912" and another called "Farm" was used for recording farm activities, personal notes and philosophy. The following recording is from this book.

> April 8, 1912
> This book is for farm, garden, surroundings etc. to keep records of the work done at home and on the ranch, thoughts about home and ranch life, aspirations and plans.
>
> Mr. Neilson is plowing and clearing the desert claim (10 acres of it) for crops. On the 31st of March we planted about 1/2 acre of Nebraska Dent Corn up in the canyon which opens into the Upper Flat. It was an almost unheard of time to plant corn but it is warm there and there is, I think, little danger from frost so I have confidence it will come out all right.

The same day we put in about 1/4 acre of potatoes on the upper part of the lower flat. Mr. Neilson now has all the middle piece plowed, all that I expect to plow this year.

Mr. Neilson (Pete) was Clarence Neilson's uncle and had his own team and wagon. He worked only for my father and had no connection with the work at the quarry.

Eventually almost all of the upper and lower flats were planted to corn, potatoes, onions and sugar cane. A number of trees including cherry, apple, basswood, elm, oak and mulberry trees were planted in addition. A small vegetable garden containing lettuce, radishes, peas, corn and melons was planted near the log house. Water came from a small reservoir in Ainge Gulch.

The next big undertaking was to build a reservoir at Orchid Spring and divert the water out of the deep gulch by pipeline to irrigate the lower flat. The reservoir was started May 24th; on June 6th Pete Neilson made a camp at Orchid Spring and started living there; on the 12th of June a 1 1/2 inch pipeline was installed and the first irrigating was done The following entries are from Father's "Farm" journal:

June 28, 1912

We are now getting past all the bridges on the desert claim. Several things I thought could be done but I didn't know just how. The reservoirs are made, except the one at Orchid Spring is small and not finished. We have demonstrated that there is plenty of water to water our crops.

July 13, 1912

Mrs. Douglass and I went up to McHugh Gulch evening before last. Mr. Neilson has now got water on all the pieces on the lower flat.

The onions and potatoes are growing but will be only about half a stand.

The corn that the wind did not destroy on the upper flat is growing nicely. . . . It is beginning to tassel out. I am learning a good deal about farming in the country. I did not realize things would grow with so little water. We probably have not had five inches of rain.

August 12, 1912

Returned day before yesterday from a nearly two weeks trip to the Uinta Mts. Went to get pictures of hunting and fishing scenes and for rest and outing. Succeeded, as I had intended to do, in throwing off nearly all care and worry about the quarry and ranch. Our prospect of a crop seems to grow more remote all the time. Night before last cattle got into the corn on the upper flat or rather the corn got into them. It was doing finely and without a drop of irrigation. Mr. Neilson says there were ears large enough for roasting. There seemed to be the best prospect of a crop but they harvested it completely

It was my own fault too, and I have no one else to blame. It would not have taken much fence but when there was time to build one it slipped my mind. I did not follow the principal that Mr. Neal told me that the first thing in ranching is good fences. In money the crop would not have brought me much but the experiment would have been of great value. It is doubtful if I could have proved up this fall and now it is more doubtful.

Work on the ranch was now drawing to a close for the year and was soon to be interrupted for a time, but the interest in its development continued unabated. Final proof was made on the homestead October 4, 1912. The potatoes were dug on December 17th with the results being somewhat disappointing though a considerable amount of potatoes was harvested.

On December 29th mention was made that there had been little time to write in the book on account of being extremely busy shipping specimens. Some 187,000 lbs. of dinosaur bones had recently been sent and were in Dragon, Utah. Contained in this shipment of bone was the last of the skeleton of No. 1, the brontosaurus, the original discovery. At the Carnegie Museum in Pittsburgh all efforts were being concentrated on this skeleton in preparation for mounting. Consequently when the last bones were on their way, Father was requested to return to Pittsburgh, for a time at least, to help with the preparation for mounting. Though he was not especially thrilled with the idea, he obliged.

He started putting things in order for at least a year's absence. On January 25, 1913 he filed application for patent on the Carnegie Museum Placer. The filing of this placer claim, which covered the dinosaur quarry, later caused this area to be set aside as a national monument. His sentiments regarding his leaving for a time are intimated in a diary entry.

> February 23, 1913
>
> I have got to leave these scenes and this country which have grown so dear to me. I did not want to go, but felt if I were obliged to, it would be a good thing for me in many ways. I can hardly bear to stay here and see the land unimproved. But I have not the money to invest so it is probably better to go away for a while. Then too, we will probably make arrangements with Mr. Kay, letting him have our household goods and other things which they want, and taking work for it. This will hold what we have done and other improvements can be made. I am especially anxious to have the land above the road that goes west from the house, plowed several times so it will be ready for crops or trees. I want to see orchard along this ledge.

⌘ ⌘ ⌘

32

MY BOYHOOD DAYS:
Early Memories at Dinosaur

My first clear memories were at Dinosaur during the summer of 1912. I remember a few things about the camp in the gulch below the quarry quite vividly; the tents, the sheep wagon where the men stayed, the horse stable, the garden and flowers, especially the hollyhocks and large sunflowers along the path.

I remember going to the quarry with my father and watching the men chiseling away at the bones on the face of the cliff, mixing plaster with water in a bucket, dipping burlap strips into it and wrapping them around the bones. I too discovered and excavated from a quarry of my own on a small clay knoll across the gulch from the garden. There I would dig and find imaginary brontosaurus, diplodocus, stegosaurus and many other species of dinosaurs. Although just four years old, I knew names and peculiarities of practically all the dinosaurs found at the quarry. People laughed when I answered their questions about the prehistoric beasts, not realizing that dinosaurs were my toys and playmates. Being the only child in camp and listening daily to "dinosaur talk" filled me with scientific knowledge far beyond my years.

I do not remember the team of horses, Mary Ann and Bess, that my father rented from Mr. Goodrich and kept until he bought the white mules, but I do remember tales about Mary Ann, who used to snake lumber up the steep trail to the quarry by herself. After a load was unloaded at the top she would return to the bottom of the hill for another load all alone and thus did not require a driver.

I can remember seeing the logs piled up for the new house at the foot of the cliff, down on the new homestead, and the men starting to build it. The next thing I remember we were sleeping in the new house and walking up to camp, just after daylight, where the cooking was still being done. I remember seeing the stars in the early morning and the folks talking about Halley's comet. I called it a star with a tail on it.

One day late in the summer, when we were pretty well settled in the new house, I was playing in the yard in a new row boat my father had bought but which he had not yet launched in the river. Taft, the Black Spaniel pup my father brought home from Murray's a year or so before, was with me, as he generally was. Suddenly the idea occurred to me that I should go up to Orchid Spring and visit Pete Neilson, about a mile and three quarters hike by the road. Though the country abounded with coyotes, wild cats and rattlesnakes, I had no fear or sense of doing anything wrong. I enjoyed the walk knowing perfectly well where I was going. I arrived at Pete's camp shortly after noon as he was eating lunch. I was delighted to be there and expected Pete to give me a hearty welcome, but he didn't seem too well pleased and began questioning me. I assured him it was perfectly alright for me to come up, that my folks didn't care and I had come to stay all night. On second thought, I might stay two or three nights. When my mother missed me she became frantic. She ran to the river, then to the quarry to summon my father and the men who rushed to the river. Mother was positive I had gone there and fallen in. There was only one encouraging factor, the dog was missing too.

Jay Kay mounted one of the white mules and, while the rest were fanatically running up and down the river bank, started systematically circling the place looking for tracks. He was almost three quarters of a mile from the house when he finally picked up my tracks and followed them to Pete's. Pete, instead of going to work after lunch, hitched his team to the wagon and was about to start for the house when Jay was sighted coming up the canyon on the white mule. Jay and Pete talked a little and Jay asked me to go with him. "No," I said, "I'm going to stay all night with Pete." Jay Kay unceremoniously reached down and jerked me onto the mule behind him and we started down the canyon at a good pace. Mother was almost prostrate when we reached home. I didn't get a whipping but was not allowed to play outside by myself for a long time. Mother wasn't herself for a while after that.

Another incident that occurred not long after added to my mothers misery of trying to raise a son in the wilds of Utah. I had been playing outside but went into the house to tell Mother there was a snake in the yard. She followed me out and to her horror discovered it was a rattlesnake. Her fear was overcome by her determination to kill the threat to her young son's safety. Courageously she grabbed a shovel and attacked. I'll never forget the excitement that followed as I watched from the house where I was ordered to stay. Taft was a good snake killer but knew better than to tackle a rattlesnake. He barked furiously, the snake was rattling and striking, mother continued chopping at it with the shovel and finally succeeded in almost severing its head. She heaved a sigh of relief. Taft, thinking the snake harmless, jumped in to give it a shake but immediately yelped and jumped straight in the air. The snake, although its head was almost severed, had managed to sink its fangs into the dog's upper lip. Taft laid down in the shade of the nearby wagon, his lip and head beginning to swell dangerously. Mother got on the phone to the quarry, (recently installed due to my wanderings) to report the events. Father and one of the men came down immediately. They fastened a stick in the dog's mouth so he could not bite them and with a razor cut his lip to bleed the poison. Taft never even flinched and soon crawled under the house where it was cool remaining there for about a week scarcely moving, never eating. Each morning we expected to find him dead but finally he showed signs of improving . After a month he was well back to normal.

This same summer we made a very memorable trip to the Uinta Mountains, (the trip mentioned in the entry of August 12th, 1912). It was the annual custom of a number of the prominent families of Vernal to enjoy an outing in the Uintas during sage hen season, usually during the month of August to escape the heat. It was almost like an emigrant train going west in the early days. Although we used our buggy, the others used covered wagons and literally moved their households to the mountains. They even took along a number of milk cows so they could have fresh milk while camping. The young men and boys on horseback drove the cattle and extra horses.

Our family was invited to join the party that year, a two-day trip to Diamond Mountain in the Uintas. We all camped together in a beautifully secluded spot surrounded by pine trees and close to a large stream. During the day some hunted geese and sage hens, so plentiful in those days, while others fished for trout. My father did both and took pictures as he went. I can remember one day's kill of sage hen laid out in a solid square as large as a medium-sized room in a house. At night around big bonfires there was eating, singing, dancing and fun for all.

The eventful year of 1912 ended and we departed in March of 1913 for Pittsburgh. There we lived in our home on Hermitage Street along with my Aunt Nettie and Grandfather Douglass. Mr. and Mrs HcHugh had been renting the house in our absence and had greatly improved its appearance by planting roses, lilacs and other shrubs. Father had never gotten around to planting flowers and Mother had longed for some around her house.

After my experiences at Dinosaur, city life had little attraction for me. I do, however, remember enjoyable days spent at the Carnegie Museum strolling down the halls and viewing the many exhibits. I also enjoyed sledding on Hermitage Street during the winter months. Father soon tired of the city and began dreaming of his life in Utah.

> September 9, 1913
>
> I was thinking this morning that I wish, before we settle down in our home in Utah and make our final plans, we could make a long trip to various parts of the United States and Canada to learn all about the various modes of culture of all things raised for pleasure, beauty and profit.
>
> This would give an opportunity, also, to see different kinds of people at best advantage. I confess I like to see man at his best and what he is mentally, rather than to dissect his bad characteristics, and hold them up to the light and let them overshadow the better traits.

I used to so much wish to go to South America and Central America and I would yet like to do so especially the latter. But my interest for my own country grows so much more intense every year and I find South America is not the ideal I had pictured.

In an entry of September 17, 1913 he tells of his financial status, mortgages on property, borrowing on life insurance, etc. to raise money to develop the desert claim and homestead. The house in Pittsburgh was still free of encumbrances. Nothing more appears in the book until April, 1914.

> April 25, 1914
> Have not by any means lost interest in farm and home but haven't time to write for other things have crowded it out. We have thought of it much and have been doing something practical. Pearl is now out living on our homestead to save and preserve the place which we wish to make our home. This doesn't seem home to either of us any more. We would be tempted to sell it if we could get what we gave. With what we could get for this we could fence our land, irrigate what we wish, start our little house so we could live comfortably and add on later. I do not want to do these things for selfish ease. I think of others' material comforts. I want Pearl to be happy and contented. She, like me, has high ambitions and I want her to have a chance to fulfil them.
> But besides my wife, who is nearer to me than anyone else, I want to get my father where he can be out in the open more, and Nettie. All my life, since she has suffered, I have wished she could be happy but I have faith all will work out well. I feel pretty definite about my duty. I am pretty well settled as to what I want to do. I want to make an ideal home. I want to, at the same time, go farther and farther into the wondrous world of thought. I want to reach the highest height I can here. I feel that one cannot attain the highest

who does not make use of all the means around him, and at his disposal, in each stage of his existence.

I want to plow, to sow, to reap, to garden, to have fine animals about me, to see things grow and blossom. I want to get at the real significance of things and help my fellow man to arise and come into the light of freedom and truth. I want suitable surroundings for this. We have had hardships all our lives. I want those around me, who have borne the heat of the day, to have calm and rest at evening.

May 2, 1914

We found several months ago that we could not commute on our homestead in Utah. This was an oversight of the registration of the land office and several others. This was a disappointment; something had to be done. Pearl at once wished to go out there. Well at last it was decided that she go and attend to the business there, hold down the homestead. The only drawback was our separation. She was anxious to go and went cheerfully. She is there now and well-established in our little home. She does not have any desire to return to the city but, of course, is lonely without me.

My interests now are there. Though I am happy most of the time when my stomach does not bother me, yet museum work no longer holds my principal interest. It is the farm, subduing rough nature and making a home.

Also camping and roughing it in the wilds, studying nature, discovering and getting its buried treasures. These things are as fascinating as ever. There will always be rugged wild places there which man cannot spoil; Always solitude and grandeur.

I do not remember much of the train trip in March 1914, when mother and I returned to Utah to save the homestead, except being snowbound somewhere in Kansas. The snow was so deep we couldn't see out the train windows and the diner ran out of food. I barely remember Mack, Colorado, but I clearly remember boarding the Uinta Railroad passenger train there, the only railroad ever to enter the Uinta Basin, which consisted of a funny little steam engine and one passenger car to match. I remember the engine puffing loudly and the wheels of the engine frantically spinning at times. I remember the deep snow and the fearful heights we were ascending as we could see when looking out the windows on the right side.

I do not remember much of the stagecoach trip to Vernal except the large wagons loaded with freight, being dragged by four to six big draft horses through the deep ruts in the mud and snow. I remember Kennedy Station, the overnight stop for the stagecoach. While eating dinner that evening at a big table with the rough-looking and rough-talking freight drivers, we listened to them brag and boast among themselves about their teams of horses and how many tons they could pull up such and such a hill under certain conditions. It seemed like a continuous argument that could get violent at any time. Judge McConnell, a prominent citizen from Vernal and friend of my parents, was on the stage. He was a huge, white-haired, well-dressed man and was ever present to help my mother, for which she was very grateful.

I will never forget how happy I was to be back in our log house in Utah and to be with my dog Taft again. Mr. Jay Kay had taken good care of the place while we were gone. Mother and I stayed alone in the cabin until August when Father returned to join us and commence the development of the long-dreamed-of ranch.

⌘ ⌘ ⌘

33

OUR HOME IN THE WILDERNESS:
Developing a Ranch

Several long entries in my father's diary while he was still in Pittsburgh tell of his plans for the ranch and home. In August he writes of the good news, what he was hoping for, he was to be sent west again. By the beginning of September he was back in Utah and brought his father and sister with him.

> September 2, 1914 - on ranch in Utah again
> At last I am where I have longed to be and where I intend to spend a greater part of the rest of my life. I am far away from sights, sounds, suffering and degeneracy of a great city. I am out where the air is pure and sweet; where we have fresh food and something of freedom and independence.
> We came in the dry hot season and things do not look so attractive as prospects for a crop are concerned yet it is good to live in the fresh fine atmosphere, to have a home of ones own, to have fresh milk, cream, eggs, vegetables.
> There are many difficulties, of course, with which we have to contend. I am beginning war on the rabbits, and we have rats, chipmunks, prairie dogs, mice to contend with.

The development of the ranch was now started in earnest. During the fall of 1914 a portion of the homestead near the house was fenced and the new house was started on the bank of the Green River by the Lone Tree. This structure was being built of hewn stone from the nearby cliffs and was of excellent construction, done by Mr. Olson, a stone mason from Vernal. Two milk cows and calves, a team of mares, a saddle horse, chickens and pigs had been purchased. Land was plowed and harrowed for planting in the spring and one plot was planted to fall wheat.

There seemed to be no certainty, at this time, that the work would continue at the quarry or that Father would be connected with Carnegie Museum for long. He had apparently made up his mind to stay in Utah no matter what happened as was noted.

> November, 1914
> If we keep on finding such good things at the quarry we may work it nearly all winter. Then we are supposed to open up the mammal quarry and do some more searching in the Uinta Formation ... If I could work for the museum until August I could probably get something to do here, if I need to.

The work at the quarry continued almost without interruption, however, because no suitable stopping place could be found. Each year more and more discoveries were made and more interesting and valuable material was uncovered, plastered and shipped to the Carnegie Museum.

The improvements on the ranch also continued. By the end of 1918 the homestead and desert claim were "proved up on" and quite adequately fenced. Reservoirs were built to catch and save the spring runoff water from Ainge Gulch and Camp Gulch (the gulch the highway now traverses). For some time, no further work was done on the new stone house but the log house was temporarily improved. A lean-to was added to the east side for a kitchen and one was also constructed on the north side, which served as two bedrooms. The log house, though not a thing of great beauty, was now quite comfortable. The main inconvenience was water. Since the spring had never been piped down to the house, as had been planned, all water was hauled from the river in oak barrels on a sled, or stone-boat, by horse. This water, although perfectly fit for human consumption, was often quite turbulent and, in some cases, could have more correctly been called mud than water.

A short distance to the east of the house, on the bank of Ainge Gulch, stables and corrals were built for the horses and cows. North of the house in the talus of the steep hill was a cellar where food and vegetables were stored for winter use and to the east of this was the ice house where ice, cut during the winter, was packed in straw for the summer. West of the cellar there was a chicken house. More cattle were acquired and together with the increase the herd totaled more than twenty head by 1918.

During the spring of 1916 Mr. Stringham lambed a flock of ewes east of our place a couple of miles and the next spring Bishop John N. Davis lambed in the same place. Both were prominent sheep men from Vernal and good friends of the family. They offered the "bum lambs" (lambs not claimed by their mothers) to us if we were willing to pick them up. It became my lot to mount Joe, the gentle old white mule, to visit the lambing grounds each morning. If there was a "bum lamb" the herder would put it in a burlap sack with holes cut for the head and legs, and hang it on my saddle. One day I came home with lambs hung all over Joe, five in all. Only some of the lambs lived but with the increase they were soon a nice start of more than fifty head.

We had our own meat which included beef, lamb, pork, chicken and turkey, and we had our own eggs. Milk cows furnished our milk and Mother made butter and cheese. Such modern equipment was added as a cream separator, a power washing machine and a grist mill with the power being furnished by a single cylinder, two-horse power gasoline engine (I had to furnish the power for the milk separator). We raised all kinds of garden vegetables which were kept in good supply most of the time.

A problem that now arose was my schooling. I was the only child of school age in that part of the country and the nearest school was seven miles away in Jensen. Mother and Father both taught me but the only children I had to play with were friends from Vernal who came out to stay at our place when their parents would permit. They were delighted to come and stay with me. The school problem didn't bother me in the least. I had my horse and saddle and my dog, the cows and sheep to care for, and all the wild country to explore when I could sneak away. I also had a 22-caliber rifle of my own. What more could a boy ask for? I visited the quarry frequently and watched the progress of unearthing the dinosaurs. People who visited us thought it terrible that I should have to be way out there in the wilderness where it was so lonesome and with no other children for companions. I couldn't understand what they were talking about.

The school problem was later solved when the William Schaefermeyer family took up a homestead on the adjoining land to the east, built a log house, and moved in. Mr. Schaefermeyer worked at the quarry and the family had four children of school age. Since my mother was a qualified school teacher it was soon decided to establish a school. The Boan family, living on upper Brush Creek in an isolated area, persuaded my mother to board and teach their two daughters making a total of seven children of school age.

The first winter school was taught in a tent adjacent to our house. The next year, with the assistance of Mr. Schaefermeyer, who was a carpenter by trade, a log school house was erected a short distance to the east of our house. Uintah County furnished desks, blackboards, books, paper, pencils and other supplies. School was taught by my mother in this log school house for about six years.

The log school house brings vivid memories to my mind of an incident that took place not long after its completion. It involved the last of the only two whippings my father ever gave me. It was a cold winter morning and after he had built the fires in the cook and heating stoves and I was up and dressed, he asked me to go to the school house and get the fire shovel which had been left there the night before. He needed it badly to take up the ashes and asked me to hurry. I started out in good faith, but for perhaps a year or more my mind had become more and more easily distracted from what my father told me to do. He had been very patient and had warned me many times about this.

On my way to the school house I passed one of our corrals where the hired man was involved in the nasty job of trying to teach the calf of one of our milk cows to drink milk from a bucket. I stopped for a moment to observe the undertaking and became completely engrossed in the whole procedure. I do not know how long I stood there by the corral fence but suddenly, without hearing or seeing a thing, I realized something had me by the nape of the neck and the seat of the pants and I was headed toward the school house at a terrific pace, my feet just touching the ground now and then. This strange force behind me caused me to enter the school house and after closing the door bounced me around the room like a rubber ball for some time then stood me up and said, "The next time I send you for something you get it." My father said no more but took the fire shovel and returned to the house. The suddenness of the whole thing made the shock terrific. I was mortified!! When he left I slumped into a desk and stayed for a long time thinking the whole thing over. Try as I would I couldn't feel too hard toward him. My errand running became much more efficient from that time on and I am sure the lesson was worth as much as a year of school. I have thought since that it would have been well if he had taught me a few more things in this manner but that wasn't his approved method of teaching and it never happened again.

My principle job at that time was the care of the sheep and cattle, a job I loved. The cattle were little trouble but the sheep became a problem because of predatory animals. The country abounded with coyotes, wild cats and even mountain lions that were lurking around waiting to make a meal of our sheep, chickens and turkeys. If the sheep were left in the hills by themselves for any length of time, or allowed to scatter, there was trouble. Caring for the sheep was one of my biggest jobs, and with my horse, dog and gun we waged continual war on the predators but with little success. They must have been watching my every movement from behind the rocks for it seemed the moment my back was turned their exploitations began. A big portion of the losses occurred in the spring and fall when I was attending school during the days.

One evening after school another incident occurred that gave my mother gray hairs and hastened the end of the sheep raising. When I started to look for the sheep I noticed six head on top of the high cliff to the northeast in sight of the house. I also noticed one lone sheep standing on a ledge below them. The rock above the little ledge was practically vertical and so smooth the lone ewe could not get back up with the others. I scaled the cliff like a "human fly", as I was accustomed to doing, and upon reaching the top decided to bring the whole bunch down to the ledge and on down the face of the cliff. This seemed easier than trying to get the lone sheep up to the others. Everything worked fine except in forcing the other sheep down onto the ledge two of them lost their balance and hurtled over the vertical cliff and were dashed to their death on the rocks 100 feet below. I slid down onto the ledge with the sheep, and as mother watched from the house she could imagine me losing my balance as the sheep had done. This was almost too much for her. The sheep had to go. They were traded for cattle in the spring of 1919.

Life was one continual round of pleasure for me at this time and I especially enjoyed exploring the Green River. I was ten years old and I could hardly remember first learning to swim yet I was not allowed to go to the river alone. I was about as honest as one could be about everything except the river. I would promise by all that was holy that I would not go near it and next thing I knew I would be there. It seemed the Green River had some strange magnetic field that attracted me to it. I would start from the house on my horse in the opposite direction from the river, and after making a big circle would end up on its banks, miles from home or any other person. One of my favorite sanctuaries near the river was at the mouth of Split Mountain Canyon where the Monument Campground is now located. As soon as I arrived there I would strip off my clothes, swim across and play for a time on the other side, then swim back and perhaps swim across and back again just for fun. Although the river looked innocent it was dangerous, even for good swimmers, and few years passed that someone was not drowned between our place and Jensen.

By 1918 the country was quite changed from that hot day in August, 1909 when my father discover the dinosaur bones. Up until the time the camp was made in Camp Gulch there had been no habitation on the north side of the river between the Neal Ranch and Split Mountain Canyon. The only habitation on the south side of the river in the vicinity, was the Thorne Ranch a couple of miles up stream.

Now, besides our place on the north side of the river, there was the Schaefermeyer family and there was always a family staying at the first camp site in Camp Gulch. Directly to the south of us, across the river, Jay Kay had taken up a homestead and south of him was the Johnson homestead. There were now five families living in the vicinity of the quarry.

The little community at Dinosaur was pretty much self-sustained and, except for occasional visits to Jensen and Vernal for supplies, we seldom left the area. All got along well together and enjoyed picnics and other social gatherings. During the horrible influenza epidemic of 1918 none of the families at Dinosaur contracted it, although several of our good friends in Jensen and Vernal died. Trips to town were cut to a minimum and when they were made those going wore gauze masks.

My father was influential in organizing a community club or betterment group that functioned for some time and seemed to be enjoyed by all who attended. Its members included families from most of the farms between our place and Jensen. It was quite well organized, having its elected officers and appointed committees. They met once a month on Sunday during the summer months, exchanged ideas on farming, and stock raising. Father gave scientific lectures and there were discussions on literature and other subjects for those interested. The place of meeting was alternated among the members and after the meeting a big pot luck dinner was enjoyed.

By a vote of the members the name officially chosen from several that had been suggested was the Lone Tree Betterment Society. Such was the way of life in the little community of Dinosaur in those days. Today the country appears almost as it did when my father arrived, except for the paved highway leading to the monument and the cars traveling it. At this point the dream of a ranch seemed to have a good chance of becoming a reality but who could tell what the future held?

⌘ ⌘ ⌘

34

FINANCIAL DISASTER:
A Lost Dream

By the fall of 1918 the ranch in Utah had progressed considerably but there was still much to be desired. A good portion of the land between the cliffs and the river, below the mouth of Ainge Gulch, was now under cultivation, or at least had been planted to some crop. The crops were meager, however, because of insufficient water, and winter feed for the cattle herd, which had grown considerably, was inadequate. Hay had to be purchased every year. The reservoirs proved to be undependable as a source of water for irrigation. Although they captured a considerable amount of water during the spring and sometimes during summer showers and cloudbursts, the water was so laden with sediment that it filled the reservoirs with mud. More often, however, the water came down with such force that it washed out the diversion dams and ditches and never reached the reservoirs.

Things looked brighter for the future. Cattle were a good price and seemed to have a promising future and we had a nice little start. Father, for a long time, had fancied purebred Shorthorn cattle and when Mr. Graham from Colorado wintered his purebred herd on the Neal Ranch, Dad became more interested and could not resist acquiring a few. Four cows and one bull were purchased at an average price of $200.00 per head. This was a lot of money but he felt they would pay for themselves in the long run and would help build up the herd.

The one thing most needed now was water for irrigation. It became apparent that the only reliable source of water for farming, except Orchid Spring which would irrigate only about ten acres of land, was the Green River. My father had been studying the possibilities of this source for some time and while in Pittsburgh in 1914 had even sent to Montgomery Ward and Sears Roebuck for estimates on pumping plants. He found them to be very expensive. It was decided that the pumping plant was the only real solution and it was ordered. To finance it the house and lots in Pittsburgh were sold. Installation was started in 1918 to be completed during the spring of 1919 in time for summer irrigation.

The year of 1919 was eagerly anticipated as the big year, the first year we would have a decent crop. Early in the spring the ground was prepared and crops planted; construction continued on the pumping plant located a little over a mile up river on a high bank. The water was lifted over sixty feet by a six-inch centrifugal pump driven by a twenty-five horse power single-cylinder stationary engine. The pump discharged the water into a ditch approximately two miles long that delivered the water by gravity to the farm land. The ditch itself was quite a project as it crossed several large washes and flumes had to be constructed.

Everything went pretty much as planned with the exception of the weather. The spring of 1919 came in hot and dry with practically no rain. No one suspected that it would continue that way through the summer. In an ordinary year there was plenty of spring moisture to bring crops up and get them well started before irrigation was necessary. This spring there wasn't even enough rain to bring the crops up and water was needed immediately. The pumping plant was rushed to completion but because of unforeseen trouble with the pump and ditch the water was too late for most of the seeds. Some fair corn was raised but there was little hay for the cattle. Even summer feed became scarce as the range land burned under the unrelenting sun. With the spring calf crop and the purebred cattle purchased we had a nice little herd now. They survived the summer drought but were thin and weak to go into winter.

The climax came with another blow delivered by nature. On top of the hot dry, spring, summer and fall, an early winter came in November with heavy snows that did not melt until the next spring. Ordinarily cattle could get along in the hills until the first of the year but during the winter of 1918 they needed feed more than a month early. Hay was scarce in the country and the price soon began to skyrocket. An average price for good hay was usually $8.00 to $10.00 per ton. Before the winter was over Father paid $50.00 per ton for hay of such poor quality that cattle less starved than ours would not have looked at it. Even with the poor feed, deep snow and very cold weather there were few losses through the winter but when spring finally came and there was plenty of green feed they died like flies. They would walk into a mud hole simply because it was down hill and when pulled out would walk right back in again. We lost about a third of the herd that winter and spring. The purebred cattle died first.

This was a hard year for a good part of the West, as anyone who lived there and was connected with livestock at that time remembers. Some sheep and cattle men were completely wiped out and all felt it in some way. The Douglass ranch never recovered from this shock, though the pumping plant was used the following season and some fair crops were raised. The recession which followed World War I had reduced the price of cattle so they were worth but a fraction of what they had been the previous year. Our cows were not worth the amount Dad had paid for the musty hay he fed them the previous winter. A government loan was negotiated on the homestead and most of the accumulated debts were paid off but it was impossible to continue the development of the ranch, although the idea was not given up completely.

One thing not seriously affected by this hard and trying year was the dinosaur quarry. It continued to produce remarkable specimens of Jurassic Dinosaurs in great variety. Since resuming work at the quarry in 1914 the west end had been opened up and a complete spinal column of a diplodocus, from the neck to the tip of the tail, a length of some 75 feet, was removed. The last vertebrae of the tail was about the size of a man's little finger. A new and deeper cut was excavated in the west end. While this excavation was in progress a nearly disastrous incident occurred. The upper sandstone layer from which the bones of diplodocus had been removed slid in filling the cut and burying everything. Fortunately the slide happened at night rather than in daylight hours while the men were at work. With little warning, it would surely have killed them. Following this excavation a deeper cut was undertaken in the east end of the quarry and work was being conducted there in 1919. More perfect skulls were found. The following diary entry mentions one of them.

> Wednesday, December 17, 1919
> Found a skull which belongs with the new skeleton No. 333, skull and lower jaw apparently articulated and perhaps articulated with neck.

Although the skeleton of the original discovery, the brontosaurus which now stands mounted beside diplodocus in the Carnegie Museum is very spectacular, probably the most outstanding discovery of the quarry was the one mentioned above. This was the skeleton of Camarasaurus, a small dinosaur some seventeen feet long, which turned out to be one of the most complete skeletons of a dinosaur ever found. It is one of the outstanding attractions in the hall of paleontology in the Carnegie Museum, where it is mounted as a slab mount in the exact and almost life like position in which it was found. Each skull discovered gave reason for a celebration. These celebrations, known as "skull feasts," were picnics or dinners given by my father for all employees and their families.

This is a fitting time to mention the employees, all locally recruited, who helped with excavating, uncovering and preparation of bones for shipment to the Carnegie Museum. They were an earnest, hard-working and reputable group and although the wages paid were small the work was steady throughout summer and winter. An interesting thought which occurs to me is, the majority of the long-term employees stemmed from Clarence Neilson, the first man, or boy, hired by my father. Clarence, whose father was John Neilson, was from a large family. His brothers Jessie and Orvil were employees at Dinosaur some time later. Jay Kay, who went to work the year following the discovery, was the brother-in-law of Clarence. Jay's younger brother, Roy Kay (later Dr. J. Leroy Kay who continued to work for Carnegie Museum after my father severed his relations with the institution), and father John Kay, were both employed at the quarry for a number of years. Two more of the early employees,

Golden and Ace York, were brothers-in-law of Roy Kay and Linas York, their father, was a long-time employee also. Joe Ainge, Mr. McHugh and Mr. Schuler, early employees, had no connection with the others. Mr. Schaefermeyer was the son-in-law of William Neal.

An interesting summer was spent in Eastern Utah and Western Colorado just prior to Carnegie's suspension of work at the quarry. Excavation at the quarry was temporarily put on hold, a house truck purchased and a camp outfit assembled. A party consisting of my father and mother, Roy Kay and myself made a trip into the far reaches of this general area in search of fossils of all ages and kinds. This trip was one Father had long wished to make and it yielded many valuable specimens including plants, leaves, flowers, insects and the skeleton of a fossil bird from the Green River Shales. Fossil mammals were collected from the Wasatch Formation and considerable time was spent in Western Colorado searching the Brown's Park Formation for material that might determine its geological age. We camped out all summer and it was a memorable experience in the wilderness for me.

No further work was done on the ranch after 1921, though the idea was not abandoned and the land was held for sometime. The homestead eventually reverted back to the government but the desert claim was kept clear and is still in possession of the Douglass family. (Editor's note: by the time this book is completed the 80 acre desert claim may be a part of Dinosaur National Monument as the family is working on a land purchase/trade DDI). The new stone house on the banks of the Green River was never completed. We continued to live in the log house until 1923 when doctors insisted that my mother move to a lower altitude. Her health had not been robust since the winter or 1909 and there had been several recurrences of her heart problems, one being of a very serious nature. This "spell" required most of one summer under the Lone Tree to recuperate. A walled-up tent with screened sides was constructed in the cool shade of the old tree and with complete rest she finally mended.

During the fall of 1923 it was decided that mother and I should go to California. The last of the cattle (about 35 remained) were sold at the depression price of $30.00 per head to finance the move.

Included in the sale was the prize shorthorn bull that had been purchased for $300.00 in 1918. Father remained in Utah to finish the work at the quarry, which was finally drawing to a close.

After the Carnegie Museum ceased operations at the quarry the National Museum removed a dinosaur skeleton which is now a mounted exhibit in Washington D.C. The last work at the site, the removal of two partial skeletons by the University of Utah in 1924, was done under my father's direction. The quarry had been a National Monument for several years but as soon as paleontological activities ceased, it was just a place on top of a hill.

The failure and disappointment on the ranch was a bitter experience for my father that led to near desperation and later despondency but it did not last for long. He had no thought of leaving Utah nor did his faith in the future of the Uinta Basin and Dinosaur National Monument, or his love for the wild rugged country diminish. He terminated his connections with the Carnegie Museum soon after they stopped operations at the quarry refusing repeatedly to return to work in the East. He could not bear to think of spending his time at the monotonous laboratories when there was so much to interest him in the West. The Uinta Basin was a new, undeveloped country and he wanted to be in on its development. He had been studying its geology for approximately fifteen years and the more he studied it the more convinced he became of its possibilities. To these interests he now turned his energies.

⌘ ⌘ ⌘

35

A NATIONAL MONUMENT:
Dreams of a Natural Museum

The dream of dinosaur bones worked out in relief at the site of the discovery was in my father's mind at the time Dinosaur National Monument was established in 1915 but even before that he had dreams of an unusual museum. Visions were apparently forming in his mind not too long after it became evident that the discovery, made during the summer of 1909, was not just a few bones appearing on the surface but the burial ground which had preserved amazingly complete skeletons of the Jurassic Era. I remember in later years how he worked to interest people, and the government, in developing the monument so that it might be an important attraction to tourists and a place of scientific study for those interested in paleontology, geology and nature. However, I did not discover until I began transcribing some of his diaries that he had entertained the idea for so long. An entry from one of his books entitled, "Uinta Geology" touches on the subject.

> August 14, 1912
>
> Have just returned from my trip to the mountains where I had mental, if not physical rest, and I came back with a keener interest in our work here. Have too, just received the Director's Annual Report of the Carnegie Museum and that encourages one. It shows that we are doing an important work, and it is having its effect, and others see it. There is surely a fascination in discovery and pleasure and encouragement in success and we certainly have been successful in uncovering great scientific treasures. What is still uncovered spurs us on.

These things being big and in large quantities attract the attention of people and the interest of those who have means necessary for carrying on the work, and it would seem that we are only in the beginning of this great discovery and research. With trained men and money we might make one of the great exhibits of the world and perhaps we will be permitted to do it. We may be dreaming and that is alright; dreams come before reality.

But wouldn't it be great to have a new museum built on museum principles and have one great hall for dinosaurs! Mounted skeletons in position of life, skeletons mounted on tables for demonstration and scientific study, restorations of beasts life-sized, restorations in native haunts, restorations showing the circumstances and tragedies of their deaths and burials, restorations or pictures showing their resurrection—how are the dead raised up?

Of all the exhibits, that of the animals and their surroundings in life, and a representation of their skeletons in the position in which they were buried would be most interesting. The former is long and difficult and requires all of man's best knowledge and imagination, to ever arrive at approximate results.

An entry made in February 1913 touches on the effort to acquire the land surrounding the Dinosaur Quarry for the Carnegie Museum.

Feb 3, 1913
We are still at the Dinosaur Ranch. Returned from shipping Jan 16. I arrived in Vernal the night of the 11th but had a good deal to attend to and was hindered by not getting check books. Had to return to Vernal the 26th to attend to matters leading to the patenting on Carnegie Museum Placer Claim and other business. Came home the 28th.

The effort to acquire the land by bringing a placer mining claim to patent continued. The case was unusual. The excavation of fossils was not deemed a true placer or lode mine by the government. An attorney was engaged by the Carnegie Museum and much evidence was presented in support of it being a placer deposit but in 1915 the area was suddenly set aside as a National Monument and thus the matter was settled.

My father's reactions to this and some of his ideas regarding its development are recorded in a 1915 diary. At this time he was out in the rugged, arid, badland country southwest of Vernal, where an interesting discovery of Eocene mammals had been made. He had taken two of his men, Jay Kay and Jessie Neilson, with him. They had established a camp and were excavating in preparation for shipping the fossils to the Carnegie Museum. From this quarry came the complete skeleton of a titanothere (Dolichorhinus Longiceps, SAB) which, mounted as a slab mount, is one of the outstanding exhibits in the gallery of fossil mammals in the Carnegie Museum. (Editor's note: Several skeletons of this mammal were collected and one is on display at the Utah Field House of Natural History State Park in Vernal, SAB)

> October 29, 1915
>
> As I am out in the desert where there are no dwellers but the transient sheep herder, I sometimes almost wish I could have his job, for a time at least. If I could choose my surroundings; what an opportunity to develop the mind.
>
> Three years ago I passed the half century mark. In one way it seems that a large part of my life was wasted, yet it has all been experience.
>
> While I am out here, I am especially favored. I do the cooking, oversee the work and usually put in my time but I have not my home duties and the leisure I can use.
>
> I have at last written Dr. Holland telling him my wishes regarding staying here. I do not know what the effect will be. It is reported in the papers that our Dinosaur Quarry has been proclaimed a National Monument. I do not know what the outcome will be but it will affect the museum and us personally. It looks as if there were going to be something doing

here and it looks as if I ought to stand as good a chance as anybody. It is a great quarry and it could be made a great attraction to sightseers and those who are seeking knowledge.

There are parks with natural curiosities, geysers, petrified forests, etc. Here is one of the greatest curiosities in the world - a burying ground of huge prehistoric beasts of a long distant age. It is a combination of fortunate circumstances that they have been buried, preserved, and again unveiled to us. How appropriate that they, or part of them, be exposed in relief as they were buried, to show the tragedy of their death and to reveal something of their lives and surroundings. How appropriate to build a fair sized building over them to protect them, to have this a thing of substantial beauty modeled after nature, to have it large enough to contain related fossils and other curiosities, geological sections, explanatory descriptions, pictures, paintings to represent scenes in the age in which they lived, a library with books throwing light on the geology of the region; anything to attract in the right direction, to interest, to help to appreciate nature and her wonderful ways!

I cannot think that this means any ill to us though there may be a limit to taking out bones. It may be just what we long have wished. The quarry here near White River is exceptionally good so far. I think I can safely say the best thing found in the Uinta.

The reports from home are that the alfalfa, wheat and barley are looking fine. I am glad they are getting a good start.

Almost three years after my father's death in 1931 an article was published in the December 13, 1934 issue of the Vernal Express, concerning what he had envisioned had transpired at Dinosaur National Monument, after he had been absent from the Uinta basin for several years. The article was given to the newspaper by my mother who stated that it was written some fifteen years previous. As the dry year of 1919 is mentioned, it must have been written about 1920. Following is the article, in part, taken from the original manuscript.

VISION OF A NATIONAL MUSEUM
by Earl Douglass

As the train descended the grade from the divide on the west side of the Uinta Basin, we were wondering how the country would look after the changes of which we had heard various rumors.

We were somewhat startled at the appearance of the Ashley Valley. When we left, much of the land was fast being ruined by an injudicious use of water and some had suggested introducing water cress, rice and frogs. . . . but the drought, culminating about the year 1919, though hard to bear, had taught a valuable lesson. Vernal was a beautiful town, had a healthy, normal growth, and held an enviable position as an intelligent center.

The principal center of interest for the tourists was the Dinosaur Museum and Park, which was about twelve miles east of Vernal, near the Green River, and for this place we were bound. We could have taken the trip from Salt Lake City in an airplane, but we wished to examine the country more thoroughly so we took the fast train. We arrived at Dinosaur, a station near the monument, at about noon.

There was an airplane standing ready, near the station. This was loaded with a party of tourists and sightseers. It started eastward and then circled to the north. We could now see the long stone building on top of the cliff. This was the Monument Hotel. It was a beautiful but substantial structure and seemed in almost perfect harmony with its surroundings.

We alighted on a leveled place north of the hotel. The plot of ground surrounding the building was kept green and fresh by the use of water pumped from the Green River. A wide, covered porch, nearly surrounded the building, and there were many comfortable seats where people could read, rest and view the great variety of surrounding scenery. Dinner was ready, and such a dinner can be eaten more easily than described. ... One thing we want to mention before we forget it, there was a large up-to-date hotel in Vernal. This would surprise the oldest inhabitants, had they lived to see it.

But to return to a most pleasant topic, this dinner at the hotel. The table was loaded with good things; tender, juicy meats, the best of vegetables were brought to the table in transparent, covered dishes, hot from the oven and steam pressure cooked. We asked the proprietor where he could obtain such excellent fruits and vegetables in such perfect condition. "Look out the window," he said pointing to the southward. We looked and where before was a shad scale flat, there were plots of orchard and fields of vegetables. We did not sit apart at small tables, sullenly gorging ourselves in solitude, but at large tables, and we helped each other, got acquainted, and were sociable. It did not take us long to realize that we were among people of unusual experience and intelligence.

After we were well-filled, and therefore good natured, and had talked and rested a while, we started off on well-made paths to the Dinosaur Museum. When we reached the peak or ridge just south of the old quarry, we beheld a sight that startled us. Extending eastward and westward, partly hidden by the cliffs, was a long building, its eastern and western portions were solidly built, spacious and well-lighted. We entered the door on the east side. Here, well-arranged in departments, were rocks and minerals, relics of the aboriginal inhabitants of the country, shells, and other fossils from fresh-water and ancient seas, fossil plants, insects, fishes and reptiles. On the walls were paintings of scenes in the different geological ages from the Carboniferous to the Age of Dinosaurs. By looking out the east window one sees exposures of carboniferous rocks near the center of the uplift of Split Mountain Canyon and the upturned ledges of rocks, of more recent date, until the beds of the upper Jurassic are reached in the vicinity of the museum. The pictures on the wall helped one to see in imagination the scenes and conditions under which this vast series of rocks had been deposited and we could have sat for hours gazing at these paintings until we imagined that we lived, as unencumbered spirits, amid these strange scenes in those long, warm, dreamy ages.

There was one exhibit in the middle of the large room which was of special interest to visitors. In the early days of the quarry, nearly every interested visitor wished to know how the bones of theses giant animals became embedded in these sandstone ledges at the very summit of this high ridge. Well, here was a plaster model of the Uinta Mountains and Basin with different colored layers representing the different formations. This was represented as cut through and the edges of the strata were nearly flat. This showed the region before the upheaval of the Uinta Mountains. There was another larger model of the mountains after the upheaval and as they appear now with the higher strata removed and the upturned edges of the formations exposed. Above this, suspended from the ceiling, was a model exactly fitting this. This showed the vast amount of rock, including the top of the range, miles in thickness, which had been removed by erosion. This was let down by pulleys so we could see how the Uinta Mountains and Basin would appear if there had been no erosion. The more we studied this we were told, the better would we understand the geology of the region if we took the "flying" trip in an airship.

I could have spent days in this room, but I saw people continually going into the next and I was anxious to see what was there. A glimpse through the large opening startled me. It was huge bones and skeletons, which seemed to almost cover the sloping north wall. As we approached we heard various exclamations of surprise and wonder as, "Can it be that it is real?". It was difficult to enter this apartment as the sight seemed to over-awe nearly everyone. We are naturally awed by the past; what would be our feelings if we could stand in some ancient Egyptian temple or palace, as it was thousands of years ago, and gaze on even a mummified form of one of the old pharaohs, or talk with Abraham or Alexander or Buddha or Jesus of Nazareth! The lips of these giants of mind and heart have long been silent but their influence and words live. Behind them man had long considered time as a blank, but there are records so old that when we learn to read them it makes us almost consider ourselves personal acquaintances of Homer and Socrates.

In the space on the wall, where there were no bones, we noticed an inscription from one of the world's oldest books. Carlyle called it the greatest and we agree with him. We thought it was very appropriate and fitting:

"But ask now the beasts and they
    shall teach thee;
And the fowls of the air, and
    they shall tell thee;
Or speak to the earth, and it shall
    teach thee;
And the fishes of the sea, and
    they shall decree unto thee".
        Job 12:7-8

Men agree now, perhaps unwillingly, and are following this counsel, and in places the dark mists of the past are rolling away and revealing a wonderful past; could it be that this marvelous display of bones was lying in the relative position in which they were buried millions of years ago? Yes it is true. We were soon trying to reconstruct in our minds these scenes of ancient days. At last we noticed that people were sitting on seats on the opposite side, and gazing in admiration at the wall over our heads.

We went to the other side and sat down. We looked up and saw a great painting. The painting was one hundred feet or more in length and represented a great river flowing through strange and unfamiliar scenes. Down this river were floating carcasses of dinosaurs, large and small. In one place near where the river found it's way into a great marshy lake, the carcasses were stranded and even being covered with sand and mud. Well, we had to confess it all seemed much like a dream, when suddenly there was a dark shadow, and looking upward we saw a dragon-like creature with wings, coming toward us. We tried to escape. As the beast seized at us we beheld one of those fierce flying reptiles with hideous, open jaws. I awoke from my reverie and my wife held me by the arm. "You seem to be dreaming again," she said.

I have not time to tell about the west room, of the fossil fish, shells insects, plants and mammals from the Tertiary deposits which we could see on the broken plane stretching to the blue horizon far to the southward and westward. I have not time to tell of our trip in an airship over Blue Mountain and Split Mountain Canyon, of our pictures in the wild parks in the Uinta Mountains, among the evergreens and poplar groves and beautiful wild flowers, of our visits to various oil regions, badlands and gilsonite mines. We spent one of the pleasantest months of our lives in this region.

We will say this in closing. We have visited eleven of the national parks and monuments in our country, and each has its peculiar sights and pleasures. There is freedom and good fellowship everywhere, but Dinosaur Monument is especially attractive to those who are intellectually inclined, and we never have been in any place where there were such intellectual treats and where we met so many people we can never forget. It is especially attractive to geologists, and the registers show names of these and hundreds of other scientists from all over the world. Our Congress used to be niggardly in its appropriations for parks, but since a movement to have our school children see the world as well as books, and trees and parks have become a part of the school curriculum, funds have not been lacking to make these national parks and monuments of the highest value to all people.

A letter written by my father to Dr. Walcott, Secretary of the Smithsonian Institute, in 1923, near the close of the excavations, is here quoted in part:

> "I hope that the Government, for the benefit of science and the people, will uncover a large area, leave the bones and skeletons in relief and house them. It would make one of the most astounding and instructive sights imaginable."

After the University of Utah finished removing its material in 1924 and transported it to Salt Lake City, dinosaur activity, which had been almost continuous for fifteen years, became completely dormant. Deep cuts had been excavated into the hill from east and west on almost the same level and the prize dinosaur material removed. This left a section in the center, some twenty feet or more deep and around 150 feet long, untouched which could be quite easily exposed. This was the area my father had dreamed might someday become a natural museum with skeletons exposed in relief in the position in which they were buried. When the work was discontinued few bones were left exposed and no one knew for sure that dinosaur skeletons were buried in this area, although it seemed very likely as the three contiguous sides had been very productive. I don't believe Father ever doubted this area contained as good or better material than what had already been taken out.

Years later, when work actually started again, the thought entered my mind many times, was it possible this might be a barren space and after it was excavated and exposed, it would be found there were no bones; nothing to see but clean grey sandstone? To my delight, this was not the case.

The years following the last excavations were discouraging years for my father regarding development at the monument. Although he tried in every way to interest people and the government in its development, little enthusiasm was encountered. He continued his geological research and study of the Uinta Basin and Utah, endeavoring to develop its resources. A couple of years were spent in Salt Lake City preparing the dinosaur specimens at the University of Utah. One of these specimens was an excellent skeleton of Allosaurus, the huge carnivorous dinosaur of that age, with a perfect skull. Later a home was purchased in Salt Lake City and his remaining days were spent there. The depression years of the 30's came and the idea of anything being done at Dinosaur National Monument became more and more remote.

My father died early in January, 1931 and during the ensuing years progress at the quarry continued to be slow. Eventually a transient labor camp was established at our original camp site in Camp Gulch and some preliminary excavating was done with the transient labor. Later a temporary building, a sort of visitor center, was erected a few feet from the site of the little imaginary quarry I had worked as a boy. During this period the name of Douglass was seldom mentioned and there seemed to be little concern about what had previously transpired or who had discovered the dinosaurs. The log house and the school house on the homestead stood for a number of years but eventually tumbled down and were burned leaving little evidence of Dinosaur Ranch.

The beautiful and little-known canyons of the Green and Yampa Rivers were added to Dinosaur National Monument in 1938. This increased the area of the monument many, many times, but contributed little toward its immediate development. Not until controversy over the proposed construction of the Echo Park Dam reached a high pitch, in the early 1950's did anything really happen.

A new era began to dawn. People began to wonder why this huge area had been set aside as a national monument. The greater part of it was so inaccessible it couldn't be seen, and a dam could not be built in it without establishing a precedent that would be injurious to other parks and monuments. Also, it was supposed to abound with the remains of huge prehistoric dinosaurs yet there were none to be seen. Why wasn't something done about it?

By 1954 things started to move and the Dinosaur Visitor Center, as it now stands, was dedicated in 1958.** If my father were alive I am sure this would give him some satisfaction and I believe he would feel that all his dreams and effort were not entirely in vain.

**Unfortunately, this visitor center was built on an unstable formation and had to be closed in 2006. According to Mary Risser,

Current Park Superintendent at Dinosaur National Monument, "The Quarry Visitor Center was constructed on expansive soils, and the first hints of problems emerged even before the construction was complete. Cracks in the parking lot began to appear in November 1957, and during the first year of operation, staff detected disquieting vibrations in the upper gallery. Over the years, the National Park Service has conducted major rehabilitation actions on the building. In 1989 work was done to anchor the existing steel roof and visitor gallery deck beams to the masonry pilasters along the south wall.

The NPS commissioned a formal monitoring program, and the first of a series of four observation trips took place between May 8 and May 11, 2006. A detailed inspection identified some previously unknown conditions that presented serious life, safety, and health hazards. Because of these concerns, park management made the difficult decision to close the building on July 12, 2006 rather than continue to put park visitors and employees at risk.

The rehabilitation of the Quarry Visitor Center is one of the projects to be funded through the American Recovery and Reinvestment Act of 2009. Construction is anticipated in spring 2010 and will last between 1 and1-1/2 years."

Ms. Risser has been very diligent in her efforts to put DNM at the top of the list for important government funding. DDI

⌘ ⌘ ⌘

36

A LAND OF UNDEVELOPED RESOURCES:
There Must be Oil

Not long after my father first entered the Uinta Basin and the great geological structure began to unfold to him, he started to ponder the presence of unusual deposits of hydrocarbons and their source. He was aware of the gilsonite mines at Dragon, the end of the Uinta Railroad. The railroad had been built over a hazardous mountain pass for the sole purpose of transporting this black shiny product, used in the manufacture of varnishes, paints, tires and batteries, from the Uinta Basin to the nearest main-line railroad at Mack, Colorado. It is estimated that about 95% of the known gilsonite in the world occurs in the Uinta Basin. Father soon discovered other hydrocarbons and the asphalt sands that are all residues of oil. In addition there were the vast deposits of oil shale in western Colorado and eastern Utah. Everywhere there were indications of oil having been present at one time. Had it all evaporated leaving only the residues or were there still reservoirs deep beneath the surface waiting to be pierced by the drill?

At that time a majority of the oil geologists and major oil companies seemed doubtful about the presence of oil. The general structure of the big basin, between the Uinta Mountains on the north and the Book Cliffs on the south, was a great syncline, or trough, running in an east-west direction. Few structures, or domes, which are natural traps for oil, were present. As time went on Father became more and more convinced that oil was present and that domes were not necessary in the Uinta Basin. He reasoned that the oil had migrated by gravity to certain areas and there it should be found. He did not hesitate to openly state his views, which did not conform to orthodox oil geology of that time, and his ideas were quite often either ignored or not given much credit. He was respected, however, and was seldom confronted openly with skepticism. I think they thought the "professor" was dreaming and perhaps getting a little out of line with his recognized field of paleontology.

Father's diaries and other writings dwell considerably on his interest in this subject and are here partially recorded. One of the first times oil was mentioned was in 1914 while he was in Pittsburgh working on the Brontosaurus.

> March 9, 1914
>
> Mr. Morgan was here last Friday and wished me to look up some matter concerning possible occurrences of petroleum in Western Ohio. This got me interested in oil, as I wish to figure out the oil problem in the Uinta Basin. I intend to make the study of the geology of the Uinta Basin a specialty.

Mention is later made in his notebook entitled, "Geology of the Uinta Basin," of his interest and study of the possibility of this region, and his ideas concerning the proper approach to the development of mineral deposits. He states:

> "I wish in this book to collect data and begin to get in shape a paper on the Uinta Basin geology. I may not be as able as some to make all sorts of theories plausible to those who don't know much about the matter but we need reliable data. There is only one thing that commands unlimited respect and confidence and that is truth.
>
> Notwithstanding the fact that all over the Rocky Mountain region there are prospect holes, which represent hard labor, and mines are caving in, and mills falling to decay, and blasted golden hopes are everywhere. I do not believe that mining if, rightly carried on, is more of a chance than most other things.
>
> If men would use cool common sense and intelligence and figure out the problem in all its phases, the failures instead of the successes would be the exception.

> The oil proposition is another which is of much importance to the Uinta Basin. The hydrocarbon deposits are probably the largest in the world. It is naturally supposed that these depots are the residue from the natural distillation of petroleum. I know of no other theory at present. What an immense amount of petroleum there once must have been in the Uinta Basin and vicinity. Is the oil nearly all gone or is there still beneath the surface an immense amount of oil unevaporated? If so where is it? In what strata? How can one best strike it? It does not seem at all probable that the oil is nearly all exhausted. But its source and its present location should be carefully studied by experts. "

Again on page 113 of the same book he comments on the peculiarities and possibilities of the Uinta Basin.

> "Do you realize that the Uinta Basin stands by itself? There is no other country like it in topography, climate, geology on other conditions. Do you know no other country has hydrocarbons like those of this basin. No other formation in the world has the remains of fossil mammals like those of the Uinta deposits, no other region has yielded such an abundance of nearly complete remains of the great reptiles of the past.
>
> Though the Uinta Basin is of so much interest from so many points of view yet it is unknown even to geologists. I hope that before long it will be possible for men, competent to do so, to more thoroughly explore the basin in the interest of its development, and to solve the problems which are unsolved concerning the oil, gilsonite, asphalt and other products."

In 1918 Dr. Marion Franke moved to Vernal as a practicing physician. He and Father were soon acquainted and became close friends. Dr. Franke became interested in the geology and resources of the basin and they enjoyed many pleasant trips together. Father's

1921 diary mentions one of these trips in which the principle interest was oil.

> Saturday, January 15, 1921
> Dr. Franke came with a car and after putting in a few provisions, camping utensils etc., he and I started for a little trip to study geology, look for fossils, and to study the oil problem.
>
> Friday, January 21, 1921
> Dr. Franke and I rode down to the Siddoway Reservoir and Mr. Lash, Elmer Grubb and DeWitt Lash came down with truck. Elmer, DeWitt and I went over and posted a notice on the claim reserved for us.
> I returned more convinced than ever that there is oil in the basin and where I supposed and that it will be found when one goes at it right.

This was in the vicinity of the present Red Wash Oil Field. An article in the Salt Lake Mining Review, August 30, 1925 entitled "Some Thoughts of the Present and Future of Petroleum," by Dr. Earl Douglass, quite clearly stated his ideas concerning the possibilities of oil in the Uinta Basin. It also delineated the general trend of thinking of others on the same subject at that time.

> "From what I have observed in parts of the Rocky Mountain region, correlated with a study of conditions as reported elsewhere by geologists, I have been forced to disbelieve in some of the theories and practices now in vogue in the search for oil in new fields. The field which I have studied most and in which I have spent much of my time - in Eastern Utah and Western Colorado - is peculiarly fitted for the study of nearly all phases of the problems of the origin and subsequent behavior of petroleum. I have also had the advantage of talking with many oil scouts, being with them in the field and learning just what they were after, and what they had been sent out to find. I must confess I have heard little about

tracing out the hiding place of the oil, but I have heard much about searching for domes.

The outstanding fact about oil and water are that they are migratory, they "seek their level." In working its way beneath the earth in porous rocks oil will behave much as water, but as oil is lighter than water it will take its place on top of water. This is the basic truth at the bottom of the so called "anticline" (dome) theory. It will get into domes if they are in the right position on the drainage area in or near the bottoms of great structural troughs. If there is no water it will be in the bottom of the incline (trough) if there is nothing else to stop it before it gets there.

But some will say that the anticlinal theory has worked very successfully and "the proof of the pudding is in the eating." We are naturally prone to look only on records of success; but with this as with narrow one sided theories, it has its dark side. It has undoubtedly led away from rich discoveries instead of leading to them.

But the point which I wish to make is the fact that closed structures (domes) occupy but a very insignificant part of this great area. The general structure is such that it is an impossibility that more than a small fraction of the oil could get into these small isolated domes.

It has been admitted that there is oil but the structure is not favorable. What this means has been interpreted by the actions of scores who look at a few of the oil showings and then start out to look for "a dome." I have traveled thousands of miles in this tertiary basin and I have never yet seen a typical dome but I certainly would never ask for a more ideal structure for the accumulation of oil.

In what I have said I do not wish to criticize my fellow geologists or prospectors. They are employed by others and they know what they are sent out to find. They realize how thankless their pains if they report anything else."

In closing, his views are expressed concerning the future of petroleum and the general belief in those days that the oil reserves would soon be exhausted and soon there would be no more gasoline.

> "No one will deny that the consumption of petroleum is increasing at an alarming rate, especially in the United States. But that we have reached the peak of the production of liquid petroleum in this country, I certainly believe that we have no grounds for assuming. When one type of structure, and that a minor one, has been persistently sought and drilled with the exclusion to all others; when it has been considered sufficient to study and map these and to practically or completely ignore the structure of the whole geological unit, it seems it is too early to make anything like a reliable estimate of the amount of oil still in the ground.
>
> Personally my studies have indicated no reason for pessimism, but I see every reason for some optimism. The human side is the only one that gives me grave concern."

The effort to prove the existence of oil in the Uinta Basin continued. After doing considerable geological work with various oil companies and other interests, all the time trying to encourage the development of oil in the basin, he accepted the job of preparing for exhibit the dinosaur skeletons which the University of Utah had recovered from the Dinosaur Quarry. His expenses while doing the oil promoting were excessive and the remuneration sporadic. The job with the university at least meant he would have a steady income. The time spent at the University of Utah was anything but pleasant for my father however, for while he was working with hammer and chisel 'reliefing' the bones of a huge carnivorous dinosaur which he had uncovered, his contemporary and competitive geologist at the university was glorifying himself. This man was promoting remarkable geological achievements, including the dinosaur material, that he claimed to have had brought to the State of Utah.

The time spent in geological studies of the Uinta Basin eventually paid off to a certain extent. The Gilson Asphaltum Company, which controlled the greater part of the gilsonite deposits, was faced with tax problems and needed the geology of the gilsonite veins worked out in order to estimate the remaining reserves. Due to his long experience and special interest in the area he was delegated this job. This work was done during the years of 1928 and 1929 and was of extreme interest to him as well as being remunerative in comparison to his previous endeavors.

During this time, together with some of his acquaintances, he organized the Douglass Oil and Hydrocarbon Company, which functioned primarily in the capacity of obtaining oil prospecting permits and in trying to interest capital investors in the development of the Uinta Basin. In 1929, after having a restraining order placed on some oil prospecting permits by the government, he wrote a letter to United States Senator from Utah, Don B. Colton, who he knew personally, and also to Senator Wm. H. King. The letter to the Senator Don B. Colton is here quoted in part.

> Cullen Hotel; Salt Lake City, Utah
> April 24, 1929
> Hon. Don B. Colton
> Washington, D.C.
> Dear Sir:
>
> I know you are interested in the oil situation at the present time, and especially as it affects Utah. I wish to give you some data which may help in trying to get a square deal for Utah and other mountain states.
>
> I have been studying the geological conditions in Utah, and especially in the Uinta Basin, since 1908. The general result is that all the facts point to the presence of oil. I have been doing all in my power to call attention to the facts and have published articles in oil and mining journals, showing the conditions, but until recently there has been no consistent effort to prospect by sane geological methods.
>
> New interest has recently been aroused in a region where geological conditions show that oil strata can be struck at a moderate depth.
>
> I had taken a prospecting permit, and on my recommendation, others had taken permits in the region.

If prospecting, which seems now on its way in the right direction, is permitted to go on, Utah's ability to produce oil may be tested. I have myself done something like $50,000 worth of geological work in the Uinta Basin because I was interested and wanted to see the truth win out.

Undoubtedly the motive of this order, to issue no more permits, was all right but it was done without knowledge of the conditions. Even if it could be done on a legal technicality, that would make no difference. It would create uncertainty, loss of faith and would paralyze business just the same.

Very sincerely yours,
Earl Douglass

A telegram from Senator Colton, dated May 3, 1929, in answer to the letter read as follows:

Secretary of the Interior has issued an order greatly modifying terms of previous order (stop) will be printed in full in the Salt Lake Tribune tomorrow morning and believe it relieves your situation both as to applications and permits. Don B. Colton

The interest and activity in the search for oil in the Uinta Basin was of short duration. The awful depression of the thirties was soon to arrive and bring most activities to a standstill. It took many more years and the urgent necessity for petroleum, produced by World War II, to finally prove the existence of oil in commercial quantities.

Although the oil discoveries in the Uinta Basin have not been spectacular, Uinta County alone, of which Vernal is the County Seat, now produces over one half million barrels of oil and nearly two billion cubic feet of gas per month. The major production of these is not from geological structures known as domes. (Editor's note: From the Utah Oil & Gas web site I've obtained this data: Utah totals: Oil (BBLs) 21,998,360; Gas (MCF) 441,650,015 Of this data, I know that Newfield is drilling near Myton - 5,313,917 BBLs of oil, Kerr McGee has 643,012 BBL oil and 113,284,007 MCF in the Natural Buttes area (south and east of Ouray). Others are certainly in this area, but I can't say which for certain. At least 1/4 of the state production is from the Uinta Basin. SAB)

This seems a fitting place to record my father's poem entitled the Geological Man.

> The Geological Man
> by Earl Douglass
> Tired of climbing and hungry and gaunt,
> In a canyon with rock-ribbed sides
> I saw a cabin, and straightaway thought,
> "I'll see who there abides."
>
> On an old, hewn bench by the side of his shack,
> I saw a gray old man
> With a quid in his mouth and a bend in his back,
> and a face with an ancient tan.
>
> "Good afternoon," he said to me,
> "How are you? just sit down.
> You're tired, I guess." His whiskers were streaked
> With gray and tobacco-brown.
>
> He looked at my face and then at the ground,
> And then at my face again.
> "From your looks and your dress I judge you're one
> Of those geological men."
>
> "Are you a rock-hound hunting ore
> and running it down like us?
> Or do you study the rocks in the books
> Of some theoretical cuss?"
>
> "I have studied both", I said to him
> "And have traveled by land and sea,
> To see how they're made and now these rocks
> Are a great revelation to me."

Then he said to me, "I have read these rocks,
And many a tale they've told
Of things that happened years ago,
And the way they hide their gold".

"You must be tired and hungry," he said,
"Just stop and your stomach fill.
Walk in and have some coffee with me."
I said, "You bet I will."

And many rocky stories he told,
And I told many to him,
Until my cup had been emptied thrice
And I was filled to the brim.

He spoke of a "lot of learned fools"
That study from books alone,
Well learned in things that never were,
Without a thought of their own.

"Some learn a little, or think they do,
And imagine they know it all.
They pride themselves on their mighty brains;
But only develop their Gall."

"And then that old, age-learned man
Woke up to themes sublime
When told of the thoughts he had caught from the rocks,
Those hoary records of time.

"Some think we're fossils when we dwell
On the past and the world that's dead:
But only the people that look behind
Have any future ahead."

"A man is a slave when all his thoughts
He takes from his fellow man;
And he has never a good sound mind
Who buys his brains in a can."

"My life alone has left me free
From those who think in packs;
And I do not trust my destiny
To cranks and crack-brained quacks.

"But I sometimes think that we have lived
When these rocks were sand and mud,
When life was in an earlier stage
And thoughts were in the bud.

"Sometimes in thoughts and dreams I view
The spirit of things to be;
And though there is little we know, I guess
That we'll be there to see.

"But man should live with nature", he said,
"And look at things as they are,
And look on the earth as his native land,
And not some far off star.

"And study the past, the future to know,
And find on earth his place,
And not aspire to be an ape,
Or live too high in space.

"We're only beginning to know the world;
There's little truth we've hit;
Our thoughts must be worked over again;
Our books are yet to be writ.

"We should live our lives and not the lives
Of reptile or parasite,
For this is the age of study and thought,
And not of the trilobite.

"A jelly fish shouldn't figure much
In a mammalian stage,
Or microscopic brains prevail
In a psychozoical age."

I shook his hand with warm adieu
As he smiled in the light grown dim;
And I often think of his kindly heart,
His coffee, his thoughts - and him.

⌘ ⌘ ⌘

37

PHILOSOPHY

While transcribing diaries and going through my father's writings, I have picked out a few things that speak of his philosophy of life. During the years spent at the Dinosaur Quarry he had several books in which he wrote concurrently but apparently when he wished to record something in a hurry he used the book that was closest to him at the time. The following is taken from a book that has no title other than what appears in the first entry.

> August 6, 1913
> I began a new thing this morning. Thought of it a minute or two ago, in its present form, yet it has been gradually developing. Thought to use this book only for "inspired thoughts," "psychic thoughts" etc.
>
> August 7, 1913
> I think of the past but do not, as a rule, really see it as it was. Have risen above some of the circumstances which held me. Had my troubles then; was really a cowering slave; slave to men, to religion, ideas of others, to circumstances. Am yet undoubtedly in bondage to some things which I hardly realize now, but probably will later. I am, I think more hopeful now. Life seems larger and fuller. It seems freer. I can satisfy my cravings for higher intellectual and spiritual things. I have a chance to investigate the things which are most interesting to me.
> I can't always find what I want to read, but I can try to write it, and this whether it leads to success or partial failure, either satisfies me, or tires me and for a time paralyzes this

longing. But I do find more satisfaction and pleasure in reading than ever before perhaps. But I find the ground is not covered.

Many of the great principles of human thought and action have been expressed and acted from the dawn of history. But there is a new age and new conditions. It is beginning to be seen by a few that the mind, like all other things, is vastly more than we dreamed. We have only begun to find it out. In its present state it is not, as a rule, large enough to comprehend it's own greatness. But when the mind is free and turned loose with nature it runs wild like a colt turned loose in a big pasture.

August 8, 1913

Printing presses, thousands of them are grinding all over the country, on things to a great extent not for higher, better tastes. On the other hand tending to degrade. The great vice of the age is insincerity. Wealth, we say, is the great vice of our people, but if we were more sincere we would put a more just estimate on money. I am certainly sorely in need of money. But it is not, to me, the greatest thing, and therefore the thought of it and the craving for it, occupy but little of my time.

I was thinking, when I was coming to the museum. What opportunities there are for man! What a world of possibilities! What a mind to make them his! But he is lost. Lost in the fog of creeds. Lost in following blind guides too narrow to see the world as it is. So many thrive by prejudice and greed. But are they to blame? The leaders are no more to blame than the followers. If we think we ought to serve man more than the God in our own minds and hearts, we need a leader. If we do not use our own minds instead of those of others it is best we have a leader and a guide.

The following poem is scribbled on the page opposite the entry dated August 6, 1913:

When minds are open thoughts come in
From many a deeply hidden spring.
Hidden from eyes which will not see
Shut out from minds that are not free.
Ah, many an inspiration waits
Through wasted lives for open gates.
And many thoughts undreamed before
Will enter at some opened door.
And many an accident has shown
Light's glimpse of ways to worlds unknown.
Forgotten keys have opened bars
To ways that end but with the stars.
Why should the boundless growing mind
By its own self remain confined,
While its great craving is for light
And its strong impulse toward the right?
Ah, not in vain the soul will rise,
To seek its home in earth and skies.
Each highest thought's a child of worth
Which has a right to healthful birth.
To wholesome food and atmosphere
Unwarped by pride, undwarfed by fear.
The mind that wills not to be free
The slave of other slaves shall be.
A life of bondage, thankless age,
and death without its heritage.

The following was written while in Pittsburgh. It is from the book "To Write What I Please." Mr. Prentice, whom he mentions, was the invertebrate paleontologist at the Carnegie Museum.

June 1914

What better time for poetic thought than in the morning after a sleep out of doors? Especially if one sleeps out in the woods or wilds.

If man from the first could have written books, what a literature! If more men had, each one, gotten his thoughts more directly from nature, what a world of beauty and originality! But, light, new thought, and advance, came so slowly that the man who had a great new thought, or way to life, was considered to have had a divine revelation and he was worshiped.

Thoughts, it is true, come slowly through books but I have wondered if the best and most poetic thoughts get into books. Some of the sweetest things seem to elude them (referring to books, DDI).

Mr. Prentice asked if I were to be banished to a lonely island and could take only six books what would they be. "Blank books," I said. But it would depend much on circumstances. I would want to take Shakespeare, The Bible, and it might be difficult to choose one or two others, and two or three big blank books for writing what I do not find in others. If there were some great works which put the charm, the truth, the beauty in nature and gave it great utterance, as Shakespeare saw it, in human nature, I would want that by all means. If there were great books which do for the sweeter things of thought, that mystic realm in which we must partly live to make life endurable, what is done for every day languidity, I would seize them. Shakespeare too has touched nature with a master's hand. Nothing uttered is more beautiful. Yet there is never enough of nature.

I am not so interested in men's commonplace thoughts and actions. I am in more deep sympathy with their progress and improvement; the kingdom of beauty and heaven within them.

So I am not at all in harmony with their materialistic ideals, their measure of all values in money. Therefore what I write is wholly out of harmony with the spirit of the age and it has no place except in the hearts of a few whose spirit rebels at the flippancy of the age. No one believes more than I in good practical everyday working common sense. The great lessons for us to learn are to not lower our ideals but from the foundation of everyday life, working for realization, building firmly to reach the heart and soul of these ideals.

When shall we learn the true essential things of life? When shall we learn, when a great teacher comes, to grasp the spirit of his message? Jesus came and taught, and the whole world has ever since been puzzling and quarreling over his message, and for 1900 years have been wholly unable to understand it. It was so plain, so simple, so common sense.

An entry in 1915, written at the time he received word that the Dinosaur Quarry had been set aside as a National Monument, touches on his philosophy. My father thought he was writing on his birthday but found out later his birthday had passed. He was out in the wilds, far from civilization collecting fossil mammals and had lost a day. This entry was recorded in the Farm Book.

October 28, 1915 – Was off a day, this was the 29th.

How great an art is writing. Is there anywhere any greater? The child can learn to write and compose. But how far from his to Shakespeare, Milton, or the great Bible writers.

I am 52 years old today. This gives me some serious thought. Of late years I do not give much time to repining and regretting the past.

We did not make this world ourselves. Things were not made after our ideals. We blunder as we do not know the exact why and wherefore. We come into existence where there is more darkness than light. We are ushered onto a scene of action where there is a scramble for more existence. With many, this is for the greater part of the time, the great concern to the end of life. Higher ideals may come for a time and longings for something better. But then there does not seem to be time for it. There are wife and children to feed, if not a struggle for competence and freedom.

We are taught by our elders, by teachers, priests, and preachers that success in life and beyond life lie in the narrow view and the narrow way. Although our religions had their origin in pure essentials, evasion of principles, substitutes for the real things, outward display, murmury, are glorified, sanctified and taught as essential things to salvation. Then, when we see the utter fallacy of these things we have been taught, there are the other extremes of atheism and senseless unthinking frivolity. Better is atheism than belief in a disreputable God. If there be a god would he not be the source of the best in man, the grandest and noblest man could conceive? How could one love, honor and respect a god in some respects inferior to himself. Better it seems to me not to have a definite belief in a personal god, and let virtue, love, truth, mercy, and justice rest on their own merits.

The following is taken from the book entitled, "Geology of the Uinta Basin." There are no dates in this book.

"Though I may not have as much interest in mere technical descriptions in geology and paleontology as I formerly had, yet as they concern humanity and the expansion of the mind, I have more interest. They are practical in helping men to subdue nature and appropriate its resources to his good in a better manner, and they give exercise to his intellect and are a source for the exercise of the imagination.

One of the great advantages of this region is the fact that men have come here to make homes. There is no more sacred name on earth than this and no man can be engaged in a nobler work than making an ideal home. The noble ambition of many a wife has year by year been crushed out by carelessness, indolence and apathy of a stark companion. And many a man's better nature, his ambitions to rise in life, have been snuffed out by a companion who could not sympathize with him or make a home, the only place where the human heart can find perfect rest and peace. Scorn your wife's love of "worthless flowers" and your scorn will some day return to you in the same or another kind of torture. Place your highest ambitions in something besides your husband, your children and you will lose out in the end.

The trouble with the human race today is that their ambitions are too low. It is better to ascend farther and farther from the beasts than, after having a mind and soul capable of the highest things, to go back to them or sink below them.

You cannot live in a far off heaven now. If there is a heaven its beginning is undoubtedly here. We hear from the greatest teacher that the kingdom of heaven is within you. This world is full of beauties when we are educated to see them. I am not a preacher but I am a man. I do not wish to preach, but my cardinal creed, that any man should be all he can be and exercise in a wholesome manner every normal faculty, should think deeply and broadly, does sometimes take me very nearly to holy ground but I am democratic and believe every man has his right there.

Do not be scared by a seemingly hard or unfamiliar word. Geology means simply the science which treats of the present structure of the earth and tries to find from the conditions of the rocks what can be known of its history. The geologist is a man like other men but perhaps with more than average interest in this particular study. His methods are simply common sense and observation. This all men should possess. He records facts as the newspaper man should do, and he should not propose theories until they seem to be in accord with all the known facts, and then they should be given as theories, and as steps to higher truths, or generalizations.

There is a tendency among those who make a specialty of some branch of human knowledge, not to overemphasize its real value, but to give it too high a selective value. There is so little known in every branch of human knowledge, and so much to know that when one gets interested in one branch, has a liking for it and lives to find out all he can about it, it seems to him to fill nearly the whole field of knowledge. I hope that people will not think that the author believes geology the only branch of knowledge worthy of man's study, though he hopes to convince many that its real importance is not usually appreciated. Perhaps there is no science better misunderstood by people in general, none in which even learned men indulge in more wild speculation. True science is sure knowledge, but as our knowledge is very much limited science as yet, covers a small field.

To those who have spent their lives in the Uinta Basin its physical features seem quite matter of fact. The stranger sees it as a weird land of chaos, of rocks, hills and buttes. The practical man sees the billions of money in gilsonite, asphalt, and other hydrocarbons, its large stock ranges and great engineering schemes which will make streams of water in the desert and make them bloom. But the man who makes the earth and its history one of his principal studies, sees more than these.

If you see a mighty bridge or aqueduct, spanning a great river, or a magnificent building, you know someone has made it and wonder at it, but you go about the rocks, over the hills, and into the mountains and take it all as a matter of course, but every day you go amid wonders that a higher power has made, and let it go at that. You do not realize that the higher power did, and is still doing these things, as truly as the builder has done, and that these methods are to a great extent traceable, as they are recorded in every rock, hill and valley, and they are for you if you will take the trouble to read them. To the admirers of the wilds and the beauty of nature they increase the charm manyfold. To the practical man who wishes to develop the minerals and agricultural resources, they are as beneficial as human nature is to a salesman. To those who are engaged in teaching or in anyway dealing with the highest faculties of humanity, they are indispensable."

The following is taken from the book with no title:
January 30, 1915

It seems strange we wish to believe in inspiration. We cannot very well disbelieve in inspiration but we sometimes doubt as to its source. We love to think that it comes from a higher source of knowledge, yet it always fails in something or if true spiritually, it fails in common everyday truth. Yet shall we throw away all aspirations; all belief in a higher destiny; all that really places man above the beast? It sometimes seems that man is a victim of a hopeless fate. He rises to sublime heights, sometimes, freed from physical bonds, he seems a spirit in heaven itself, yet his physical body and his earthly relations and environment pull him down at last. He struggles to rise and he is something higher and better, but cruel fate has offered him no refuge from pain, sorrow, disappointment, death and death to all his higher ambitions.

It is a most damnable state of affairs that this should be so. The thought is repugnant, blasting, paralyzing. The more we seek pure and higher things the more enduring pleasures they give. Physical pain and infirmity may obscure them but they rise again and again. We would keep them always did not the beast in our natures drive them out. What then should a broad minded, enlightened man of this age or any other age do? So far as I can see he should use that mind and do the best he can. He has his mind and his judgment; his aspirations to be the best he can. He should use his mind to think. There is a lot to think about.

Of the many poems written by my father, with the exception of two or three that appear in his diaries and one that I am familiar with, the dates of their writing are not known. The following poem was written in 1925 while mother and I were living in California. We were staying with Mother's cousin, Ethel Dwyer, in Huntington Beach. Father, after doing some geological work in Arizona and Texas, visited us and stayed a couple of weeks. While there he heard that a famous painting of Christ was to be exhibited in Santa Ana, California. He expressed his desire to see it and Ethel volunteered to drive him there. Mother and I went along. To my disgust, I cannot now remember either the title of the painting or the artist. I do remember it was said to have a value of $75,000. It was a lone exhibit and after we had entered the building and gazed for some time I noticed Father was at the back of the room, standing alone and motionless. After we had been ready to leave for some time he was still standing there. We finally left but no comment was made concerning the painting on the way home. For the next couple of days he acted strangely and was very quiet. He was by himself most of the time and did a good deal of writing. On about the third day he handed us the following poem to read and we could then better understand his peculiar actions.

> A Picture of Gethsemane
>   by Earl Douglass
> It was a picture on a wall–
> A painted picture– "that was all".
>
> It was a picture of despair;
> Forlorn the one in anguish there.
>
> No light was in the darkened sky;
> No flower or blade of grass was nigh.
>
> With head on arm and tangled locks,
> One lone form knelt on barren rocks.
>
> No face was seen, but attitude
> And raiment were with woe imbued.
>
> Arms, hands, and all were telling there
> The awful language of despair.
>
> O, what a type I thought in tears,
> Of all the fruitage of the years!
>
> Because of these I scarce could see--
> Had I not known Gethsemane?

Did e'er a brush with truer power
Portray a darker, sadder hour?

The painter, too, it seemed to me
Himself had known Gethsemane.

- - - - - - - - - - - -

Lone one whose hopes had been so bright
Until that last and fateful night!

Faithful you labored 'till that hour
To break the tyrant's fearful power;

Boldly pursued your daring plan–
The freedom of your fellow man.

Then came the last, despairing night
With scarce a single ray of light.

Had you believed "The Father's" power
Would save you in that trying hour?

Or trust that sympathy for man
Lay deep and sure in Nature's plan?

But no one came to set you free
From that dread hour at Calvary.

How often had the words you spake
Pierced the sad heart that swelled to break!

"My God, My God, I trusted thee!
Why hast thou now forsaken me?"

Sad were the words that Caesar said
When from his Brutus' sword he bled;

But that a man's or empire's loss;
Not a world's fate upon the cross

Did you believe man e'er would be
Worthy and able to be free?

Did you not know that deaf and blind,
The living truth we seldom find?

Even the few at our right hand
Could not your message understand.

So long as men are willing slaves,
Worship and serve designing knaves,

So long those who would set men free
Shall end in dark Gethsemane–
So long in vain seems Calvary.

I've often wondered if you knew,
After your time, what men would do?

How they would take your name, and plan
The bondage of their fellow man;

Would fill the world with shame and crime
Through all the centuries since that time;

And tyrant systems named for you
Would cancel all you tried to do–

Systems that all your precepts wrecked,
And made your words of none effect

Your precepts given to make us free
Were turned to tools of slavery.

In that dread hour was there a gleam
That made the future darker seem?

In murky skies did red light break
From martyrs burning at the stake?

And on the earth a redder flood,
Where Christians shed their brother's blood?

The banners, with your name unfurled,
Long waved above a blood-strained world.

They used your name to trap mankind,
Self-blinded leaders of the blind;

Who nothing did you bade them do,
These were the ones condemned by you.

"Great is Diana", shouts the mob
And in her name the people rob.

"Great is the Christ", and many sell
In his great name the fears of hell.

The noble man that should have stood
For all that's manly, brave and good.

How oft' the "House of God", since then,
Has been a murderous, thievish den!

And yours, the greatest of all names,
Yet used for greater lies and shame.

If these things you could then foresee,
How black, indeed, Gethsemane!

O, hero of the true and good,
Two thousand years misunderstood!

Has your cause met unending shame
By counterfeiting of your name?

Is all so dark? Are there a few
To your high aims and purpose true?

Many there are- a quiet throng
That help the world and men along

To aim for higher earthly goal
And for the freedom of the soul;

Who, in a wilderness of fears,
With love and hope have gilded tears.

These are the followers of thee;
And many know Gethsemane.

Though near the end of mortal life
Did you see far beyond the strife,

Down through the ages yet to be,
Past many a sad Gethsemane,

Glad Freedom's flag at last unfurled
Above a long-enduring world?

Will men thy potent message see–
Or the world's last Gethsemane?

- - - - - - - - - - - -

Still for some future good we grope,
Inspired by sympathy and hope.

Many will die in brutal fight
But few, like thee, will die for right.

Truth, on a cross is sanctified
And men, not truths, are crucified.

Not silenced at Gethsemane
That first clear voice to set men free.

- - - - - - - - - - - -

I know not what the truth may be,
But this, that picture said to me.

Another of my father's poems, which I believe was the last he wrote, was entitled "America Beware"

<u>America Beware</u>
By Earl Douglass
Before our times men dreamed romantic dreams
Of freedom and the brotherhood of man;
And systems formed, to make their dreams come true;
That still gave man a tyranny o'er man.
    America beware.

Before us, many times, have nations risen
To power and wealth, but built on slavery

Have risen by selfish power and precedent–
But in their fortune was their own decay.
    America beware.

See now the deserts where great empires were;
Egypt and Babylon -- where now are they?
Now in the jungles and the ocean isles,
Far flung, are relics of sublimer days.
    America beware.

The times have burst into immortal song
And soared to heights of great creative thought,
Where are the Shakespeares and the Angelos,
The Mozarts and the saviors of mankind?
What are they to us if we heed them not?
    America beware.

We pride ourselves upon material things
But steel will rust and gold cannot be life;
Cultures and cults have risen high and died;
Systems developed but to fall apart;
For these are not the things by which men live.
    America beware.

Mankind has ever worshiped noble men–
After their death – but seldom did they heed
Their purest teachings or their message learn,
Or see the warning written on the wall,
    America Beware.

No law emancipates a slavish mind.
A man is slave to what he most desires.
Freedom, like virtue, is not sold or given.
A government by men depends on men--
Rises and falls as men advance or fall.
    America Beware.

As long as system curbs the better self
And man cannot develop what he is;
As long as privilege is not for all,
How can we hope for progress that endures?
    America beware.

This reverence for false tyrannic creeds,
This lack of reverence for Eternal Truth,
This worship of the godlike men of old
And scorning of the godlike deeds today,
This resting on a name that others made
And never doing worthy things ourselves
Can never save a people from decay.
    America beware.

During most of the year of 1930 my father was troubled by a prostate gland condition which caused him considerable discomfort and was beginning to interfere with his endeavors. He was advised by his doctors that he could continue in this condition perhaps a year or two. He was also advised that by having an operation, which would be serious and possibly fatal, he might successfully pursue his work for a number of years. He did not hesitate in his decision. He wanted the operation and during the latter part of January 1931 he entered the hospital. All went well and the operation appeared to be successful until the second day when his kidneys failed to function. He died the next day, January 13, 1931.

Father's funeral services were conducted in a Salt Lake City mortuary by Dr. Elmer I. Goshen of the First Congregational Church. Dr. Goshen, an outstanding and controversial minister in Salt Lake City at that time, was deeply admired by my father. Dr. Goshen gave a brief review of his life and work which. He concluded by reading the following poem written by my father.

The Great Unknown
    By Earl Douglass
There is a land beyond our ken
Where mortal step has never been;

Beyond where the sun's last rays have shown
There lies the strange the vast unknown.

We live on earth with our own kind
And share the kinship of the mind,
But each, companionless and lone
Must drift away in the great Unknown.

There's nothing sure in our passing years
But pain and troubles and death and fears;
Unsettled and fleeting is all that's known
But we rest our hopes in that great Unknown

Vague pictures sometimes cross the mind
But the boundaries are undefined.
We long to see what's never shown,
To know what's hid in the great Unknown

We long for a life where sorrows are o'er,
Where questions and doubting perplex us no more;
Where life to fuller existence has grown
In the land of all hopes, in the great Unknown

We hope that the good that in this life is done
May not die forever but live on and on,
That the good seed we've scattered that never has grown
Will bear rich fruit in the great Unknown.

Are there the harvests we look for here
That were blasted in life's unfruitful year?
The sowing is here but that which is sown
May blossom and fruit in the great Unknown.

The dreams of earth, the dearest and best–
These aspirations that cannot rest–
Too sweet for earth, have they not flown
To meet their hopes in the great Unknown?

EPILOGUE

None of us, the living members of his family, ever had the privilege of knowing Earl Douglass, however, through his wife, Pearl and son, Gawin we have known him. We have heard the stories of the discovery of dinosaur bones and life on a Utah homestead from our grandmother before her death in 1955, and our father before his death in 1998. Earl Douglass is still honored by two granddaughters, six great grandchildren and eight great-great grandchildren. We are awed by his extreme dedication to science and relentless search for truth. We continue to be drawn to the West, its big skies, open spaces and rough terrain, partly because of our heritage. Earl's obsession for books and critical reading can be seen in great grandchildren and one great-great grandson is now turning to philosophy for his major in college because he read and listened to the philosopher's poems.

My father, Gawin Earl Douglass, was the only son of Earl Douglass, and admired, revered, talked about his father and lived with his poetry and writings until his own death at age 90. Grandpa died when my father was only 22 years old so Dad didn't have the advantage of knowing his father for a great many years, but when you consider his early years at Dinosaur National Monument as an only child, born when Grandpa was 46 years old, it is understandable that there was a close bond between the two. My father reveled in retelling stories of his adventurous youth lived in the wilds of the Utah canyons and I loved to listen, but I also helped type the Earl Douglass diaries when I lived at home in my early 20's so I learned about him by reading and typing as well.

After World War II had come to its conclusion, our family made a trip to Utah for the first time in my lifetime. We stopped by the tiny museum near the dinosaur quarry and Dad began to relive his childhood. He had managed to keep an 80-acre desert-claim situated within the boundaries of Dinosaur National Monument which had been taken out by my grandparents in 1912. When we returned to his boyhood home it was a great disappointment for him to see it in such sad condition. The old log cabin where he and his family had lived was gone, burned apparently by government employees who didn't want to see it looted and destroyed. The museum his father had dreamt of had not come to fruition and there had been little done to honor the memory of his father's discovery and the difficult years put in by Earl and Pearl Douglass during the excavation of the dinosaur bones.

As I reflect on both men, reading the journals of my grandfather and the writings of my father, poring over the old photographs, it is certainly nostalgic for me. Some of Grandpa's writings touch my heart as I realize the difficult life he led 100 years ago and the circumstances of that era. He agonized over what the truth really might be and how he could possibly put together his strong beliefs in the Christian faith and his obsessive desire for scientific discovery. In the late 19th and early 20th centuries it was not easy to reconcile these two ideologies and battles were being waged on many fronts. I believe in today's world there can at least be discussions on an intellectual level regarding science and religion and many scientists are also men/women of faith. My husband and I have found people in our churches and our own college campus here in Iowa (Central College, Pella, Iowa) who are willing to enter into discussions and continue to respect one another in spite of differences in belief.

My father, Gawin, saw his father's dilemma and chose to become an agnostic. He was adamant in stating that there might be a god, and undoubtedly there was a creator, but a human being could not know this creator on a personal level. When my sister and I began to question and believe (our parents insisted on dropping us off at a Sunday School near our home) many arguments arose in our family that were never resolved.

As I continue to go through my grandfather's diaries, writings, and poetry I am encouraged to find little bits of writing, on scraps of paper, that express a kind of reconciliation with his scientific study and religious upbringing. One such poem I found, handwritten on a small, blue piece of notebook paper, when I was going through one of my grandfather's files after Dad's death in 1998. In searching the internet I have not been able to find an author for this poem and there were words scratched out and others added, so I assume this was one of Grandpa's compositions, though he had not signed it.

<u>Where Christ Hath Been</u>

There was one that was out in the darkness,
And the winds were dreary and cold,
Away in the lonely desert,
A sheep astray from the fold.

His heart was lonely and cheerless,
And hopeless, he sank in defeat,
When he saw in the earth beside him,
The prints of the Savior's feet.

Another was out among the people,
To free them from error and wrong,
And long did he hear, never yielding,
The opposing voice of the throng.

And oft when he seemed defeated,
At midnight he offered a prayer,
And he knew in the cold and the darkness,
That Jesus had often been there.

As I read this poem I realize Earl Douglass was a constant seeker and even though many people of his day did not understand him he continued to seek truth and also The Truth that is spoken of in the Bible (John 14:6).

My father, Gawin, included Grandfather's poem "Hymn of the Wilderness" in Chapter 31 of this book. In this poem Grandpa states that among these rocks and hills (of Eastern Utah) "the prophets never trod. . .Yet here the Spirit dwells, divine, that spake with men in Palestine." In these words, and the entire poem, I see a belief that God is present in this wild wilderness and if men would but listen they could hear Him. The title of this book is taken from my grandfather's quote from Job 12:8 and from what I can gather of him, this is something he truly believed. When I read the entire 12th chapter of the book of Job in the Bible, I conclude that in his deepest thoughts Earl Douglass believed true wisdom comes from God and by listening to His creation we too can find that wisdom. This is what I have gained in knowing my grandfather through his writings and in editing this book.

⌘ ⌘ ⌘

INTRODUCTION TO APPENDIX

June 28, 1908 letter to Dr. Holland, Director of the Carnegie Museum in Pittsburgh, PA
Earl Douglass' 1908 Field Notebooks II and III and

In early 1995, I was contacted by representatives of the National Park Service to do a paleontological field survey of a private land holding within Dinosaur National Monument. The acreage included Orchid Draw, an area that was a 1912 desert claim held by Earl and Pearl Douglass during their years working the dinosaur quarry for the Carnegie Museum. The designation of Dinosaur National Monument in 1915 by President Woodrow Wilson included eighty acres surrounding the quarry, but not the Orchid Draw area. This project intrigued me, as I had been introduced to Gawin Douglass many years before when I worked as a naturalist at the Quarry Visitor Center. So philosophically I, as well as most American paleontologists, consider Earl Douglass our intellectual "ancestor" and see Douglass' dream of an exhibit wall of in situ dinosaur bones at the Quarry Visitor Center as the central diadem in the crown of American paleontology.

In the summer of 1995 Gawin Douglass visited Vernal, UT and gave a presentation to the Historical Society on his father, Earl Douglass. This presentation included numerous photographs of the Carnegie (Jensen) dinosaur quarry from Douglass' discovery in 1909 to the early 1920's when that phase of excavation was completed. After his talk, Evan and I met with Gawin because we were in the early stages of doing a temporary Utah Centennial exhibit for the Utah Field House of Natural History on "150 Years of Paleontology in Utah." Gawin allowed us to copy many of the photographs. After several long talks and a visit with him to Orchid Draw, he requested that we help him edit and publish this biography of Earl Douglass. While editing the manuscript, we noted that there was a gap for 1908. Only the quote from Dr. Holland described the 1908 field season. This was unusual because normally Douglass was a prolific journalist.

Later that year, we visited several museums in the eastern United States looking for early photographs of paleontology in Utah. During our visit to the Carnegie Museum in Pittsburgh we worked closely with Elizabeth (Betty) Hill, the collections manager. She gave us access to numerous Douglass papers, including letters to Holland and the 1908 JOURNAL!!!! She allowed us to photograph those documents, not Xerox them due to conservation concerns. But as it turned out, she had transcribed the journal and many letters. The June 28, 1908 letter from Douglass to Holland as well as the 1908 Douglass journal are included here.

There has been a misconception about Douglass' work in the Uinta Basin, as most writers credit him with only being there in 1909. In fact, the Carnegie Museum sent him prospecting for Eocene mammals in the Uinta Formation in 1908. O. A. Peterson had visited the area of northeastern Utah in 1893 for the American Museum of Natural History. His letters to Osborn tell of his search for Eocene mammals which he wasn't allowed to collect because his area of interest was on tribal lands. Instead, he diverted to look at the Jurassic beds north of Vernal where he was disappointed that he found no mammal material. Instead he wrote, "it proved to be one more of bone diggers disappointments: a few fragments of large Jurassic reptile. I have been looking over the Jurassic grounds for a few days but do not think it well to collect the bulky material on account of the expensive transportation to railroad" (Nov. 17, 1893 letter to Osborn). With this knowledge, Douglass and his bother-in-law, Frank Goetschius, also prospected the Mesozoic beds near the Green River in late July 1908, well before Holland's visit in early September of 1908. During that earlier visit, they found significant evidence of dinosaur material – a Diplodocus femur and a huge femoral head of another large sauropod. Evidence of these discoveries was documented with photographs. When Holland arrived, Douglass and Goetschius squired him around the area, showing him a variety of quarries where they had removed mammal, plant, and insect fossils. At Holland's urging, they took him to the Jurassic beds on the southeast side of the Green River by Split Mountain. Holland mentions in his writings that they went separate ways and shortly Douglass fired his shotgun letting Holland know that he had discovered bone. Dare we say that they set Holland up? Certainly afterward Holland was more than anxious to send Douglass back out to the Jensen area to find the animal "bigger than a barn" for Mr. Carnegie's new museum.

Our greatest thanks to the Douglass granddaughters, Mary Margaret and Diane, for allowing us to participate in this project and to finally fulfill the promise we made to their lovely father.

Sue Ann Bilbey, Ph.D., Geologist and Paleontologist
James Evan Hall, Photographer and Paleontologist
from Vernal, UT March 14, 2009

APPENDIX

June 28, 1908 letter to Dr. Holland, Director of the Carnegie Museum in Pittsburgh, PA
Earl Douglass' 1908 Field Notebooks II and III and

June 28, 1908 – Dragon, Utah
My dear Dr. Holland:

I have reported quite regularly to Mr. Stewart, but knowing your interest in the museum expeditions, I am sure that you will be glad to hear from me directly and know just what we are doing. If you still have it in mind to visit us on your return from Europe, you can get some idea of the country where you are going.

The principal part of our outfit I secured in Grand Junction, Colorado. This is a fine town in the valley of the Grand River. Near it are the most beautiful orchard lands I have ever seen. The valley is large and is nearly surrounded by high bluffs and cliffs of Cretaceous age.

I found it to be a good place to outfit, the only great difficulty being the purchase of suitable horses at a reasonable price. Horses are "way up" in the West. I looked at every horse I could find for sale. There was just one team that I felt was worth the price - a span of blacks, about 1200 or 1300 lbs. each. The price was $400. One nice gentle team in good condition but not large or any too young was offered me for $300, and the owner afterward wanted $350. A livery man offered to let me have a choice of three or four teams including harness, saddle horse and saddle at $2.75/100 per day rent and when rent amounted to price of outfit he would give bill of sale. Under the circumstances I thought that would be the best thing to do. It would give me a chance to try the horses and to see if there was anything here, so I chose one of the teams. The price was $250. They are gentle and both can be ridden, which is a great advantage. They answer our purpose very well and on the whole it may be as well as to purchase a $400 team which would not be as good "rustlers" on the range and if one died it would be a pretty heavy loss.

We drove with outfit to this place except over Baxter Pass where we put wagon and outfit on flat car and shipped it and rode the horses over. It is about the steepest wagon road and certainly the steepest railroad grade I have ever seen. It is on the Uinta RR.

We stopped a few hours near Mack on the Rio Grande and Western R.R. to examine the Jurassic beds there and found where hundreds of pounds of bones of Brontosaurus? had weathered out. We did not dig much and do not know whether it is all weathered out or not. The beds are well-exposed here and could probably be followed for 50 or 100 miles – I know not how far. It certainly looks favorable for dinosaur collecting. It is said that near Fruita, a few miles east of Mack, a party from Chicago dug out a fine skeleton of a large animal. We stopped for a few hours each at several places along the road and examined Laramie, Wasatch, and Green River beds. Found a few fossils in each. The freshwater and land mollusca are abundant on the Wasatch. Found only a fragment or two of mammalian bones but careful, persistent search would probably bring an adequate reward.

After arriving here (between White River and Green River in Utah) we searched the upper and more barren portions of the Uinta deposits for a few days but found only fragments.

In the lower Uinta or "Transition" beds we found fossils more numerous and in a better state of preservation. We found a well on the stage road and less than a half mile away a cabin owned by the Gilson Mining Company. We got consent to occupy the cabin, cleaned it up the best we could and went to housekeeping. It has three rooms, does not leak much and we find it immensely better for our work than a tent. It is centrally located and we drive out several miles in every direction. I think little or no collecting has been done here before. There is a stable for horses and we have the advantage over former collectors in many ways. We are taking advantage of these opportunities.

We get our mail at Bonanza Stage Station though our P.O. address is Dragon. The stage passes going each way (Dragon to Vernal) every day. Freight teams also pass nearly every day.

We have been out of the valley but once since we came. We were compelled to go to Vernal, a pretty little Mormon town in a beautiful valley 35 or 40 miles north of here.

The scenery along Green River where we passed it is charming and the change was delightful. We crossed the river on the ferry as there is no bridge.

Between Green River and Vernal we passed, most of the way, over Cretaceous rocks. It would take a volume to tell all the geological and other interesting features of the country.

Vernal, though over 100 miles from any regular R.R. (The Uinta was built by a private company to get out Gilsonite) is a lively little place and one can get many things as cheaply as at R.R. towns.

We returned as quickly as we could and have been here at work ever since. We are having success all the time but it is steady. Fossils are not at all abundant anywhere so far as we have been able to prospect, but we are getting something all the time. We, I think, are doing as well as any other expedition that has been here – probably better for the time spent but I usually arise at 5:00 to 5:30 a.m. and am busy until 8:00 or 9:00 p.m. My brother-in-law, Mr. Frank Goetschius, is with me. At present it would not be economical to hire another man for cook or anything else. The weight of this work would fall on me if I had six men. The excavating is not hard so far. What it requires is great care.

I am doing better work than ever before and am doing it scientifically. We will, I hope, be able to tell from just what level each specimen comes. This is of great importance here. I have now 85 numbers on record and in cases of invertebrates, fossil insects, plants, etc. each number may include 100 to 200 specimens. We have good skulls and parts of skeletons but just how the latter will turn out no one can tell until they are worked out in the laboratory.

One day I made a collection of about 150 specimens of fossil insects – beetles, bugs, flies, ticks, mosquitoes, (imago & larval), May flies, aphids etc. beautifully preserved. All are small.

As soon as we work out this region – yes, before that, there are several places to which we expect to make side trips. They are more or less promising, but in one summer we can cover thoroughly only a small spot in this vast area of Jurassic, Cretaceous and Tertiary deposits and exposures.

July 2

Since writing the above we have struck a quarry. Do not know how far it will extend but is very promising. Have worked on it a little over two days and have moved a lot of dirt. Uncovered 7 or 8 more or less nearly complete skulls of three genera of mammals, also limb and other bones. We have in it two good skulls of the Eocene representative of Elotherium, and Dinohyus, several of Eocene Titanotheres and one of the giant Uintatherium. It bids fair to be quite extensive but we have opened only a part. Will be easy working. I think you will not regret sending us here this summer.

I hear that you have received high honors in Europe and I hope you have had a pleasant, as well as profitable trip.

Your sincerely,
Earl Douglass

Earl Douglass Field Notebook II, 1908

Wednesday, July 1

Yesterday we hitched to the wagon and drove N.E. of here to make some measurements, map strata etc. and search some rocks we had not examined. Stopped at the foot of some low bluffs just above red beds at bottom of horizon B, hobbled horses, turned them loose and went up to the sandstone rim where we had stopped searching and measuring a few days before. I saw a few fragments of bone and dug in, soon my pick struck something hard and I pulled out a piece of a skull. I got the fragments of rock away and found some more pieces of bone. Put what I could back in place and began clearing away dirt and digging around. At a little distance saw some other fragments and began digging and brushing away dirt. Soon came to bone. I thought it a head of a limb bone, uncovered more and found that it was the condyle of a skull and the other was in place also a large area of bone which I took to be the postera basal part of the skull. After quite awhile Frank said, "That looks to me like the back of the skull."

I looked again, "Why of course it is." It was plain enough when one thought of it. It was undoubtedly Uintatherium but it was pointing nose straight down. I had lost hopes of getting a whole skull but this revived them. But could it be possible that a skull 3 ft. or more long was buried nose down? Well, in digging around, we came to what seemed the condyle of another skull close to it; finally in about the right place we came to some teeth, the molars badly worn, but evidently one of the Elotheres. I think it is nearly complete. Then we found bones, fragments of jaws, parts of skulls etc. Yes, Frank found the greater part of another skull, but it was broken up in digging down to get at other things. I thought it would be a great thing to get a Uintatherium skull complete from this horizon and thought it would contribute much to make this expedition a success.

Today we went over and did a lot of digging. Dug below in a long crescent and then worked toward specimens. I also worked around Uintatherium skull and it is evidently only back portion but very complete as far as it goes. Probably tomorrow we will begin to take up things. Do not know how far in the quarry will extend or whether there is much beyond what we can see but I think we will get some more. I worked until I was tired and lame and prospected awhile. Frank worked awhile longer and prospected. We found two or three things that will probably be worth saving. These are the best beds we have struck but as is usual with such, are limited.

Monday, July 7, 1908

Have been so busy since the 4th that I have not had time to write. Frank concluded not to go to town the 4th and it would be too dull lying around here so I proposed that we go over to White River, he to fish and I to look for fossil fish, insects and plants in the Green River shales. Ever since we went over there on horseback I had wanted to go. The exposures there are so excellent that I felt sure we would get something.

We arose at about 3:30 a.m. We had considerable to do and wanted to get an early start. Took provisions, sleeping tent, bedding, feed for horse, tins, etc. intending to stay 2 or 3 days. We got started at about a quarter to 7:00 I think it was. We went east or a little N.E.

on a road that goes to a place where someone has tried to ranch. We thought there would be a trail leading over to the road we wanted to strike, found a few tracks but soon lost them and picked our way for a long distance over rough ground and through sage brush. At last we struck the road and followed it to where a road branches off that goes down a ravine or canyon to White River. It seemed a long way down the canyon to the river. For some distance it is the Bridger sandstones and is not so steep-sided but when it enters the Green River it gets steeper and steeper until it is quite picturesque. The road though, is an easy one.

written July 8th

On the White River we saw cottonwoods, willows, buffalo berries, clematis, sedges, lambs quarter, rumex, green grass, etc. besides the shrubs such as sage, grease wood etc. that grow on the upland. There were a few mosquitoes there and plenty of flies, including little black gnats. There is a little fly all over the country now that is bigger than the "no-see-ums", that keep flying in front of a fellows face and sometimes in his ears. They are sometimes very annoying.

After eating a lunch I went to the lowest exposures of the rocks and worked upward looking for fossils. I did not find any for 50 or 75 feet – perhaps more. Finally I came to a place just above a cliff where a mass of rock was separated all around from the other rocks. I began working it down layer by layer. Spent the most of the afternoon there. Got perhaps 200 or more insects. They were in some of the layers all the way down as far as I went. In some they were very numerous. I sometimes found groups of 6 to 10 beetles in a place. One thing that interested me much was the pupa of a lepidopter. It is a dandy and I think Dr. Holland will be much pleased with it. Beetles are very numerous. There are some of the long-nosed beetles. Diptera are also very common. There are lots of little things that I don't know. Wish I had a good, well-illustrated entomology here. It will revive my interest in insects.

Frank came by under the cliff and reported that he had not succeeded in catching any fish. After he went to camp I looked a little above where we had been working and I found shales containing myriads of the larva of some large insect. On our way down I had found one on a little slab and thought it was the body of a large insect. There were nicely preserved insects in the same shales but much smaller. I found also what I supposed was part of the feather of a bird but it was time to quit and I went to camp.

Our tent was by a little green cottonwood grove. Nearby was a half-reclining willow with a heavy top. Beneath this was our kitchen. It was pretty warm so we did not care much for hot food and did not cook much. The first night we slept out doors in our sleeping sheet & tent. The mosquitoes bothered for awhile but it got cool and I went to sleep. This was the way we spent the 4th.

The next day was Sunday and I was up and prospecting early. Above the place where I had found the larvae the day before I found larger ones that I think are a different species. They were still more numerous in certain layers. Still higher found larger worm-like larvae of two or more kinds. Working up still higher Frank and I found more insects and fossil plants nearly to the top where the Green River shales are overlain with sandstone. I saved a large slab of the larvae and several specimens will be excellent for exhibition.

That afternoon Frank pitched the sleeping tent as it looked like rain and in the evening it was protection from the mosquitoes. The wind blew hard in the cottonwood trees, especially in the evening and they bent like grass in a gale. A bird – I never remember hearing the song before – is quite common there. I think it is a cat bird but I did not see it singing. It certainly is a great singer. It is the mocking bird here at the cabin that furnishes the music. I had only 4 plate holders so could take only 8 pictures as I did not take any extra plates along, so I had to select the pictures. Had one misfortune, I had Frank take one and he used the same plate I had last exposed. He supposed I had turned the exposed plate back. It is likely that we will go there again some time. If we do I want to get a dozen or more pictures.

The trip was a success and the work was done mostly on our own time so it was clear gain to the museum. Yesterday we did not go into the field. Frank put up mail box so we can get our mail every day. Also went to Bonanza to get bread etc. We undid all the specimens we had collected on our trip; did them up in better shape, labeled and put them on record and boxed all but two large packages which would not go in box. Filled one box with them and nailed and wired it ready to ship. Have nine boxes ready and specimens enough to fill three or four more.

Today, July 8
We went to the quarry again. We have found no more skulls that are good and the quarry does not appear to extend much further than we had prospected at first, though have found some other jaws etc. We came in at about noon today as we got out of fierce weather but am thinking of going out this evening and bringing in some things.

Sunday, July 12, 1908
I did up most of my work yesterday and determined to have today to myself. Have been busy part of the time but am too lazy to do very much. Have been writing, reading, etc. Wanted to do more writing.

We left a couple of specimens at the quarry but have cleaned out everything else in sight. Got about 20 specimens from the quarry. We may get some more.

East of the quarry Frank found a nearly complete skull of Telmatotherium, and one of a Titanotherid that I do not know. I think it must be new. It has very large heavy incisors, as heavy as canines nearly and the diastema is between P1/ & P2/. I think it is one of the most interesting things we have found. It is well that I started to map those beds. It is the only way to get things clear and straight though not a very pleasant job. I will have to correct my provision diagram of beds with mammals in record as I have some things wrong. For example, there are two separate red layers near bottom of formation instead of one and the numbers must be separated. We now have several fine fossil levels and will have some interesting results.

A man by the name of Burton, who is engaged in placer mining north of here on Green River, was here yesterday and day before. Took dinner with us yesterday. He says that there are bones of large animals in two or three places there. Also fossils in the Carboniferous. I want to go there before long and investigate. Part of them are undoubtedly Jurassic Dinosaurs. I think perhaps we may go next Saturday. There is interesting country all around here.

It is hot during the daytime now. We get up and get out into the field early and get our field work done so we do not have to be out in all the worst heat of the day. For two or three days we came in at 12:00 or 1:00 o'clock and went out again in the evening. It is dry and sandy here so it makes the heat of the sun quite intense. It is fairly comfortable in the shade. Yesterday we went down to Kennedy's Station to get butter, flour, feed etc. Borrowed an old cat and her two kittens so we have hopes at least of getting rid of the miserable rats and mice that are so exasperating. Caught 6 or 8 with traps but if anything they seemed to get worse.

Have a mail box down by the road now. Have not received the film packs I ordered 5 or 6 weeks ago so I have not for a long time had the use of my film plate camera.

Have most of the specimens boxed now; have one or two boxes to finish packing and nail up.

Sunday evening

I am on top of the hill south of the cabin. The sun. . .it's liquid light surrounded by a brilliant white ring is just going down below the long wooly line of the ridge west of me. Now it is hidden but its light brightens a space above, shines through holes in a long cloud and brightens its edges. . . In the yellowish, greenish-bluish. . .of the sky are little long oases of clouds

July 16, 1908

Monday I went over to the region of Quarry 1 and continued measurements toward the east. I am getting the thing straightened out so that I can tabulate fossils etc. Worked hard in the hot sun and came home pretty near sick.

Tuesday we drove over to a cabin southwest of here where Mr. G said there were some fossil bones. We did not find a fragment near the cabin. There are excellent gilsonite veins there; have been worked to a considerable extent.

From there we struck northward to a branch of Red Bluff Wash. In some bluish gray beds we found some bones worth saving. One specimen is an artiodactyl and consists of many fragments of skeleton, including hind feet etc.; also some lower teeth.

Wednesday we went to the same region again and Frank found hind feet fragments of limb bones and, I think, part of the skull of a small perissodactyl.

Today we went farther and struck higher beds. I found a nearly complete skull of a small animal. If it is an artiodactyl I think it is new but it may be a small carnivore. The teeth are there but not exposed so one can see shape of molars.

July 23, 1908

We have taken our trip over to Green River and are back in our cabin again. Have not been out collecting today but are trying to get things straightened up. We need rest too as it was, in some respects, a hard trip as it was so hot and the roads were rough, at least part of them.

We started Friday Morning, the 17th and got quite an early start... took along our grub box with provisions for several days, a bale of hay, some oats, picks, shovel, cameras etc. We drove to the well and watered, filled our water can, drove on the stage road to the top of the hill and then turned off the road to the eastward where, a few days before the freight teams had pulled the boilers over to the Green River nearly north of here. After a mile or two we struck an old road that the freight teams had followed. It led to the northeastward many a long mile. It seemed to be almost endless and all the way up hill. The road seemed to go clear to the Raven Ridge but we turned more to the northward on a crossroad, still continuing to ascend. At last we came to sloping cedar bluffs, the border of the bench or plateau of Uinta deposits. The road began to descend. Finally we came to a road traveling off. We supposed that it went to a spring so decided to follow it. It led a long way from the other road but we saw no spring, though we were since told there was a spring ½ mile

APPENDIX

from the road. We finally got into the valley of Cockleburr (same as Cliff Creek, I think) but no water in view. It was burning hot and we had been a long time without feeding or watering, but we pushed on still hoping to find water. Once we stopped and gave the horses some water out of the can. At last we crossed the dry channel of Cliff and another creek. The channels lie between perpendicular walls of red alluvial wash. We took two or three pictures, which ought to be good. After driving a little farther we came to a green grassy spot that was marshy. A little farther we found a marshy spring and a place where people drank but the water was warm and had the odor and taste of black mud. We managed to dig off the layer of black, stinking mud from the foot of sandstone slope, dug into sandstone and got pure, but not cold, water. After feeding the horses and eating a lunch, we went out prospecting to examine the rocks and find a place to camp. The rocks just above us formed a ridge of massive, and partly laminated sandstones.

Sunday, July 26

We crossed this ridge (above-mentioned DDI.) and came to a little valley between this and the larger dome-shaped uplift, called section ridge on Hayden's map. This little valley is caused by removal of softer strata. The rocks in a near this ridge stand at a high angle in some places nearly vertical. This massive sandstone, therefore, is a wall left after softer strata from each side has been removed by erosion. The soft rock in this little valley and skirting the west side of the uplift is red and yellow, principally red, and from certain levels springs issue and flow down toward the (Saargash?) or Cockleburr. These springs are mostly warm and are strongly impregnated with sulphur. They probably contain other minerals also. We thought we were going to get a nice cold drink after the long hot journey and Frank was going to follow one to its bubbling source to get a sparkling drink, but he was disappointed. The day, however, was getting cooler and the scenery was refreshing. After being on a desert of sage and sand for days, and rocks, here were tall grasses, sedges, red castellias in bloom, gage milkweed in bloom, buffalo berries, clematis, water cress, staphyler? and I can't remember what all. These springs flowed into the flat and formed green meadows. Much of the scenery made me think of other days and other scenes.

Earl Douglass Field Notebook III, 1908

July 1908

We found a little stream that flowed from a spring down the slope onto the meadow where the water did not smell or taste of sulphur. It was fringed with thickets of small trees and shrubs, grasses, sedges, flowers and weeds. We found an ideal place to camp and returned for our horses. I took several pictures along the little stream. That night, for the first time for a long time, the horses had plenty of green grass and water near at hand.

The next morning (Saturday, July 18) I arose at about 4:30 thinking that I would take a little walk, examine the red beds (marked Triassic by Hayden) near and soon returned to breakfast. But I struck a road, saw a ranch, Lombardy poplars, and farm buildings near and high red bluffs beyond and though not very strong I kept going. When I passed the ranch house no one seemed to be up so I went across a little valley where there were green thickets and a cultivated field to where a deep steep-sided ravine separated me from the red bluffs. I managed to get across this and began examining the red beds climbing steep slopes when it was hard to force my way up. I did not find anything again. The beds are mostly red dirt. There is some sandstone which is laminate. Toward the top of the exposure there is much hard sandstone.

After awhile I returned to the ranch and found the people up. I talked for some time with the man who is running the ranch. I do not remember his name. The ranch, I believe, belongs to a man by the name of Barret. The man who runs the ranch is middle aged. He and his wife have four or five children. I had thought what a nice place it would be near where we camped to have a little place, utilize the springs for irrigating etc. I think the slopes near the little runlets would raise excellent fruit. There is a large meadow which could be made to produce a large crop of hay. It would make a good stock ranch. The gentleman said that 160 acres next to the Barret ranch had been owned by a cattle company which had broken up and it was for sale. Above that they had bought a relinquishment of a man and it could be homesteaded. If a person could get it at a reasonable rate he could make a fine place of it.

When I returned to camp Frank was away. The sun was hot. I got something to eat and went in the shade and ate a late breakfast. After Frank returned we hitched up and went by the ranch taking several pictures. One showed grass and other vegetation as high as the wagon. Took a picture of the red beds (Triassic). The garden was by a reservoir below the house. I took a picture of squashes, corn and potatoes. It seemed good to see a garden again.

Below this ranch the valley is quite narrow and the stream where there is one cuts a deep, steep-sided gorge in red alluvial deposits between hard rocks on both sides. In one place the rivulet takes a couple of leaps down a couple of big stairs in the massive rock. There was not enough water to form a continuous stream. I tried to get some pictures of it. We stood under the upper fall and had a nice bath. I also took a picture of the gorge below the falls. (Continued July 28) When we crossed the ravine farther below we found two freighters there with teams and machinery for placer work over on Green River. Teams that had pulled up the grade came back and helped them pull up the grade. We took pictures of them.

After this we followed the road across ravines and up hills 'til we were above Green River. We saw no fossils but the rocks appeared to be of Cretaceous age, and overlay what we supposed to be Jurassic. When going along the ridge above Green River we saw a section of a dinosaur bone and supposed it had been left there by some one who had started to carry it away as a relic.

We descended the hill to the river by the placer camp. It was a dreary looking place and the wind was blowing a gale and raising the dust by the cabin and stables across the river so we did not signal for the ferry boat but went to a cottonwood grove a little way up the river. Here we found wood, shade, and shelter enough but no grass that the horses would eat. We had taken along a bale of hay however, so got along very well. Mr. Burton came to the river opposite us and asked if I were Mr. Douglass. After awhile we went to the new home he is building on this side of the river and found him. He asked if we wanted to go to where the bones were. We told him yes. I was surprised when he said that they were in the bluffs of soft rock a little distance west of the camp. He showed us one or two places where there were broken bones and then after talking awhile he returned as he had much to do. We went on hunting in the same general level. Did not find very much for a little while and I did not think the prospect was very good. Later Frank found a large femur and other bones and near the same place I found a mound about 45 ft. in diameter from which there were bones sticking out on every side. There was a huge broken femur 2 ft. across the head and 66 inches long, a fibula 4 feet long, vertebrae, part of a toe bone, another large limb bone or two etc. I think there were bones sticking out or weathered out in 12 or 14 places around the hill and one or two across a little hollow on a neighboring hill. It struck me at once that it would be an excellent place to excavate. It is soft dirt but on digging in a little it is partly hardened sand. Judging by the bones that have not weathered too much they are about perfect, hard, thoroughly petrified and complete. The conditions for collecting seem ideal. The bones may all belong to one individual but I rather think that more than one is represented. Frank had uncovered a part of a bone. I dug down about where I thought the other end ought to be, found it and it measured about 66 inches. I do not think it was as thick a bone as the one above mentioned. We were considerably elated over the prospect and I determined to report the thing without any overdrawing the picture to Dr. Holland. So the next day we went up there with the camera, searched the dinosaur beds and took pictures of the larger bones on the little mound as well as the mound itself.

When Frank ascended the steep, partly perpendicular slope to the beds, he found, at a much lower level some large bones, evidently of a pelvis but I did not think that it looked just like any I had seen. We did not find any other prospect so good as the one found the day before.

Monday we crossed the river – no Sunday afternoon, and went to a place opposite Cub Creek. I think it is where Mr. Burton had seen bones at the foot of a river cliff. We judged by his account that this would be an excellent place also.

When we struck the river bottom we found a boy on horseback. He was herding goats. He had a camp by the river and was living alone.

We went to some cliffs above where this river had cut into dark shales. Found fish remains but no dinosaur bones. The next morning we went below and found the place. The bones are in rather hard blocks of sandstone, are considerably broken and the prospect does not seem very promising. We did not know for awhile but we gave it up. We went again to the dark shales above and collected fish scales, bones, etc.

Wednesday, July 29, 1908

It was very hot in the middle of the day. In fact, we had a hot trip all through. Quite late in the afternoon we started on our return journey. West of the Green River Gorge the Triassic, Jurassic and Cretaceous rocks stand at a high angle and it is an excellent place to get a section of them also to examine the rocks for fossils. I had one film left and took a picture where the older beds stand at a high angle and the Marine Cretaceous shales come in contact with the sandstones that stand like a sloping wall. Before night we came to some ranches on the north side of Green River. I had heard that Mr. John Murray had some large bones. I heard that he had lost one leg. When I came to his ranch I saw a one-legged man on horseback with a shovel showing someone how to irrigate. I introduced myself, he gave the pony to Frank and came and rode down to the house with me.

The bone was part of the femur of a dinosaur, not extra large, probably about like Diplodocus. He got it from the hills a little distance northeast of there. We stayed there all night. He has an orchard and garden and it seemed good to see these again.

The next day we went down Green River to the stage ferry, crossed, pulled up to the spring on the hill and camped. Fed horses hay and grain and hobbled them as had not much hay. The next morning I found they had started for home, a thing they had not done before. I followed them for 3 or 4 miles and returned with them. Did not feel very good natured toward them so did not have much time coming home. Found our mail had been left at Bonanza. I was so anxious to get it on one or two accounts that I got on Barney and went after it. Received what I hoped to get, a letter from Dr. Holland and a check that helped us some. It was Wednesday that we got back. Thursday we did not go out in the field; were worn out and had enough to do here. Friday we went to Kennedy and got a load of hay and oats.

Saturday, August 1

We went down to White River to try to get some fossil fish. Mr. Gurr said that his brother had found some fine ones when digging out rock to fill when constructing road. We found no signs of fish but got some plants and insects from the Green River shale.

Sunday, August 2

I wanted to rest but there was enough work for two days, we had got so behind on account of being away. Did a washing, read some and wrote a little.

Monday, Tuesday, Wednesday we have been working down in what is called the Devils Play Ground and west of there. Get something every day, mostly bones of Artiodactyls and shells of turtles. Yesterday Frank found some black fragments of bone where I had found fragments of Artiodactyl skeleton. I dug in. Thought it a rotten limb bone but broke off three or four pieces and on examining them found that they were parts of top of skull. Today as it was so very rotten, put plaster all around. Lifted it up and rock came away showing that the lower jaw is in position. Part of one tooth showing makes it evident that it is a small Artiodactyl. This is an important find. Hope we will get more skulls. As near as I can judge, we have now about the most important collection made from this region.

But the end is not yet. I must rush this thing along as rapidly as I can as there is lots to do yet.

I. I want to make a trip to Ouray and examine the beds east of there as they, Mr. Peterson writes, have never been explored. I hear of fossil skeletons being found there.

II. I want to go east of here and examine the Wasatch. I wish also to give the Wasatch a thorough examination in other places.

III. I want to make a collection of fish from the Green River.

IV. I want to collect the dinosaur bones near Mr. Burton's place.

V. I want to examine the gray beds overlying the red Uinta Beds near Green River and east of the Dragon-Vernal road. It may be White River or Transitional to W.R.

Friday, August 7, 1908

Have been so busy lately that I have not had time to write in my journal or notes, and I have forgotten just where we were some of the days since I wrote. Sunday I took a rest or tried to rest – it was pretty hard work though. Wanted to write, and did write part of the time, but would get lazy and sleepy. Would lie down on my couch to rest and would go to sleep. Monday, I think it was, we went over northeast of here to collect fragments, odds and ends, from the fossiliferous sandstones in which the Quarry 1 was located. We found, or had found before, a good part of a skull and Frank had dug in one place and found some fragments. I dug some more and found it was a nearly complete skull of Telmatotherium. We had broken it considerably but we have taken up the rest in one block and saved the pieces. It will make a very good specimen. The nasals are complete. We worked in that region two or three days. Frank found part of a skeleton of an Artiodactyl in the shale beneath.

Thursday I think it was, we concluded to go to Kennedys Hole, go up one of the ravines and try to prospect in the lower Uinta and get some wood too (no, I guess that was Friday or Saturday) I found a mile or more west of the Dragon-Vernal road a lower jaw of Amynodon and part of a skeleton. We went over there today. The part of a skeleton consists of perhaps a half dozen vertebrae, a scapula and a pelvis, neither complete. Also, some parts of ribs. I found also a turtle that will do to go on exhibition. We have 167 specimens on record and 9 or 10 in the field. There are several places not yet searched that will almost surely yield some good things. I think we cannot complete the work here in less than a month and we might find work here for the rest of the summer.

We have been living pretty poorly for awhile. We did not have anything that tasted very good. We sent to a Mr. Haugh (?) to see if we could get vegetables. We also had an order in at the Ashley Co-op. We got that, or part of it. We got vegetables from both places so we are well supplied now. We also got some apples. I cooked some beets and potatoes, and we had some raw onions. They tasted mighty good. We will live high for awhile. We have potatoes, onions, beets, carrots, turnips, and apples.

I sent a long letter to Dr. Holland this morning telling him what we have been doing.

I think the weather on the whole is not quite so burning as it has been.

August 12, 1908

Have lost my fountain pen so have nothing but pencils to write with.

We are still having success. Saturday we went north of here a mile or so to examine some rocks that we had not thoroughly examined. I found in the bottom of a layer of sandstone the skull of a Protitotherium. The front part, including the premolars, is gone but the rest appears to be about perfect, including the side zygomatic arches and the high flat crest of the occiput. Frank found the skull of a Telmatotherium. Afterward I found part of the skull of a young Telmato. We ought to be able to find the rest. Then in shales higher I found a large part of the skull of another large animal. I don't know whether it is Protitotherium or not. It is broken and the front part gone.

Sunday we rested, or tried to. I wrote a good deal. Toward evening we went over north of here where we had found the skulls. Monday forenoon it rained. Rained the night before. In the afternoon we went down to Kennedys and got 9 bales of hay and 2 sacks of oats, also some butter. There was a big shower behind us when we started and we followed a big one home. It rained a little where we were.

Yesterday we went over to the Devils Play Ground east of the road and prospected. Found bones of a small lizard, an Artiodactyl, a carnivore, and some other things. Also found a large turtle which I will probably save.

Today we went over to the same region west of the road, pasted a large turtle that Frank had found, then went and pasted one I had found farther north, just south of Kennedy's Hole. I then went to take pictures of some… farther west. Frank got some fine negatives. Mine are not developed yet. I am trying the Ray Screen. A little south of us were sandstones of which there were…I said there ought to be something in them. We prospected for awhile and found what we long had sought, a skull of Diplcodon. It is not perfect by any means. Apparently one side of the skull will be good and will give the length, height etc. It may not be Diplacodon. If not, it is undoubtedly a new genus. The crowns of the teeth are broken off. The second molar was the largest tooth, the third the smaller. In this respect, it differs much from Telmatotherium or Titanotherium. The posterior portion of the skull of Diplacodon is unknown. The type is the maxillary with the teeth, and Hatcher found the anterior portion of a skull with the horn-cores. This is one of our interesting finds. We tried to take up the skeleton of carnivore which Frank found. It was a bad thing to handle. The matrix comes up in blocks that break where they will. The top came up and left a large part below. Part of the bones are lost but there are one or two fine feet. I think we have not spoiled anything good. I do not know whether we have any jaws or teeth or not. I hope so. Our good luck continues. What I want to find now is horses and rodents. I don't understand why we don't get more of these. We have no primates either.

August 13, 1908

This morning we got a pretty early start; got up at about 5 and started at about 7:00. It looked some like rain but I wanted to get the things we had found over in the Devils Play Ground. Was expecting to do a lot of work but we had worked but a little while at the first turtle when it began to rain and kept raining harder and harder. I concluded it had set in for a days rain or more. We could do nothing at pasting so returned but the sun was shining before we reached home. I did some other work and Frank sharpened the picks.

In the afternoon we went over N.E. of here to measure some strata as I want to get a complete section of these beds. Measured 150 feet of strata above the Fossil Quarry layer. This takes to the Green sandstone that caps the bluffs there. It takes up about to the red and gray beds, in fact there is some red in these beds.,

I found where the skeletons of two Artiodactyls, a little fellow and a larger one, had been broken up also where there were lots of foot bones of a Perissodactyl and a turtle. Frank found maxillary with P/3-P/4, M/1 & M/2 of a...Artiodactyl.

Monday, August 17

Yesterday was Sunday and I expected to have the day to myself but the lumber (300 ft) came from Vernal. After lunch we went down and got it, also some water. When we got back we unloaded the lumber. We put down enough in our living room to finish the floor as about half of it had been taken up. It is much nicer now as we do not have to run around on sleepers and have lots more room. I swept the yard and kept at work nearly all the rest of the day. Today we went over to the Devils Play Ground to get a turtle and part of a skeleton we have been trying to get for so long. Also went to Gurrs and got 100 lbs of flour, 175 lbs hay and 81 lbs oats, also engaged some butter. We are running short of funds, are considerably behind in fact, but I am expecting a letter and check from Dr. Holland every day. I expected to hear from him long ago. There is 40 dollars still due on the horses and I owe Frank nearly $30 on last months wages. I have used my money until it is nearly gone and have not made the usual payment on my home this month. But I am not worrying. I mistrust though, from the way money has been coming in, that we will be called in earlier than I expected. It will be extremely unfortunate if we are as the prospects are so excellent for a good falls work. Of course it would be easier for me at home with my wife and child but I am extremely anxious while I am here to make an unusually good collection and there is nothing now to hinder it but the money.

It has been cloudy most of the day and has sprinkled a little. Is cloudy this evening. There has been a change in the weather within the last week. Had some heavy rains last week—Thurs. & Fri. I think. It is now getting dark, and the katydids or crickets are singing in the sage and other brush along the run in front of the cabin. They sound at a little distance like frogs. There was water in the run two or three times, quite a large stream. I heard a wind coming in the bushes at a distance. Now it is here and is cool and fresh. I am writing evenings now.

Friday, August 21

Yesterday, as we were expecting some things by stage, I left Frank to attend to them while I went over northeast of here to finish measuring a section. Went to the gray bluffs formed of the strata below the red and gray beds of the Uinta. I had measured to the top of the gray bluffs. I was happy and light-hearted as my mind was occupied with hopeful aspirations and pleasant thoughts. It was a pleasant morning.

When I got over there I did not find the beds favorable for measuring as I thought they were so I stopped the horses, after I had driven to a place where we had not been before – dropped the tugs, took a pick and sack and went prospecting. Looked over a ledge of sandstone and old iron gravel and underlying clay finding nothing but some fragments of turtles. Went farther north and followed a little reef of gray & brown sandstone in red beds. I prospected a little distance on the plain and was low. Soon, near the end I found some little fragments of teeth and bones. Was tracing them up when I looked up at the sandstone ledge and saw a black spot. Examined it more closely and saw that it was the incisor teeth of a large animal. I did not know at first whether they were upper or lower. Soon I found that it was the front of a skull and so far one side that I thought they were out of place, I saw parts of under. It is a large broad skull and in a peculiar condition. I think the other side of the skull may be all there. Apparently the outer side is mostly there, as far back as the molars go. Then there is a...space weathered out and with it part of the skull. The lower part of the skull is soft and much broken. The upper part runs up into sandstone almost as hard as flint. It will be hard to save but I believe it can be done when we get plaster. Found other parts of jaws and other bones. Today I started out again to measure strata. Thought I would begin at the red beds below the lowest fossiliferous layer from which we had collected. We followed the butte mesa northeast of here to its farthest portion. Then I got lower beds but looking south I found other small... areas. Went over there, found red shales again and a little farther south, sandy loess with harder brown spots of sandstone that tends to weather out thin slabs. (This writing continues for another page but is so illegible and incomprehensible it would be a waste of my time for me to continue. If, in the future, someone needs to translate this, they may do so. Elizabeth Hill)

September 9, 1908

    Have neglected my journal for a long time so will have to write from memory. I think it was Monday, August 24 that we started to explore the Wasatch beds east of here, a trip which we had long contemplated. We went S.E. of here on the road that follows the gilsonite vein, then turned on the road going northeastward. It was a long, tiresome road. We passed some exposures of Uinta near the ridge east of here and then entered a ravine going through Green River shales. We had seen bare looking bluffs on high hills to the eastward and supposed they were Wasatch, but they proved to be Marine Cretaceous. We had passed the Wasatch before we knew it, but the exposures were not good. We then crossed a wide belt of

Laramie and then came to the Marine Cretaceous. The series from the Green River dipped at a steep angle to the westward. This is apparently the western limb of a large dome... When we got into the Creta., Marine Creta. the exposures not so great. The M. Creta. formed a line of semicircle of bluffs which surround Raven Park on three sides. The U. part of Raven Park is a great flat, somewhat undulating, cut by ravine. The rock is a dark shale which quickly disintegrates on exposure to the rain and...so one can see the original rock only when a stream has cut a bank or someone has excavated. Where we stopped for lunch I examined the shales in a bank and found fish scales. Saved them but the rock disintegrated and I lost them. Should have wrapped them at once.

In this region there is quite an oil excitement and we saw an oil derrick. We struck the river at the Rector ranch. It was a delightful change and I enjoyed it very much. A man was harvesting oats.

We forded the river. I took three pictures of the sun shining through the cottonwood grove which surrounds the house. When we got across the river we had a good camping ground. I determined to take some pictures the next day so on the 25th I crossed the river and took a lot of pictures. The house is a fine one. I suppose there is not another one as good within 45 miles. Vegetation is rank and there is a great variety of plants. I took pictures of the house, sunflowers, cattails, wheatfield, ponds, ducks, harvesting oats, etc. The hills we had crossed seemed pretty dreary but that furnished a very good background to the pictures. I met Mrs. Rector who was very friendly.

The next day we started down White River. Camped at night below the Goff Ranch where I thought we could get some fossil insects from the Green River shales. We had prospected a little along the way where the river cuts through the Laramie, Wasatch, etc. Found no good fossils. That night we were kept awake by the mosquitoes whose ancestors were buried in the high bluffs where we slept, or tried to sleep.

Dr. Holland came to see us a week ago yesterday and left yesterday. For the first two days we prospected or looked around here. Then we planned a trip over where we found the Dinosaur bones.

It was not as bad as our previous trip. The road across Dead Man's bench is however a tiresome one. When we descend the bluffs we go up and down [illegible] miles through stunted cedar. We went through the valley where Frank and I went on a previous occasion. Camped in a meadow in front of the House on the Burdett ranch. Had a very pleasant time. We stopped to examine the Laramie as we crossed it. In the morning we examined the Triassic exposures.

September 10, 1908

From here we drove over to the placer ground where they are putting in a dredge boat. We were disappointed in not finding Mr. Burton at home. We concluded to stop there hoping that he would come the next day. He did not come and so we pulled over to Mr. Murrays. He was not at home, had gone to Vernal. There was a man there and Mr. Murray & family returned before we ate dinner. We got some potatoes, apples, chickens, oats etc. there. Drove to the Uinta Ry ferry that night. Could not get across as it was late so had to camp. Had no hay but there was some grass. The horned owls hooted in the night, the coyotes serenaded us and the wild geese flew clacking over. Monday we drove over to the Dead Man's Bench getting to Kennedys at about 12:30. Ate dinner there and then drove home. Were glad to get here. I have not felt very well since my return and have had the itchy and trouble with my legs, worse than before.

September 15, 1908

The weather is getting quite comfortable. I have not been able to put in quite so good time as have not been feeling very well since our last trip. Sunday I had a good day. Study life and...geography along the run that goes by here. Started out with one pack of films. Exposed all of them and returned. Went out again in the afternoon and took 18 more pictures. I expect to pay for these myself but wanted them before I went. Can't get them after I go. Want part of them for an article for the Guide if they come out well. They ought to. I took pictures of most of the flowers in bloom in cutbanks, sage, etc.

We are now gathering in the specimens that are out, Telmatheres, Diplacodon, Proteotheres, etc. Went over N.E. to the supposed Diplacodon skull today. The front part is good where not weathered, but back is gone. Found another front part of a skull about 3 rods away, though it may not be the same thing. Neither may be Diplacodon but I do not think they are Telmatherium. Got the supposed Diplacodon from the Devils Play Ground several days ago. I think it is something new.

We may get through here this month but will have to rustle if we do. I want to get into Pittsburgh about the last week in October.

(Here Douglass' writing in Field Notebook III, 1908 ends. The rest of the notebook, a considerable number of pages, is blank. Elizabeth Hill – Carnegie Museum of Natural History).

***These journals and the letter from Douglass to Holland are printed through the Courtesy, of the Carnegie Museum of Natural History – Vertebrate Paleontology, 1990-1 from the Papers of Earl Douglass and Dinosaur National Monument

⌘ ⌘ ⌘

Made in the USA
Charleston, SC
17 April 2013